Studying Animal Languages Without Translation: An Insight from Ants

Zhanna Reznikova

Studying Animal Languages Without Translation: An Insight from Ants

Springer

Zhanna Reznikova
Institute of Systematics and Ecology of Animals of SB RAS
Novosibirsk State University
Novosibirsk
Russia

ISBN 978-3-319-44916-6 ISBN 978-3-319-44918-0 (eBook)
DOI 10.1007/978-3-319-44918-0

Library of Congress Control Number: 2016959596

Printed on acid-free paper

This Springer imprint is published by Springer Nature
The registered company is Springer International Publishing AG Switzerland
The registered company address is: Gewerbestrasse 11, 6330 Cham, Switzerland

Preface

Communication systems of animals as well as their cognitive abilities have been a matter of special interest to biologists, psychologists, linguists, and researchers in many other fields including robotics and artificial intelligence. From time immemorial, people have been dreaming about understanding animal "languages". However, decoding the function and meaning of animal communications is a notoriously difficult problem. The problem of cracking animals' codes have become especially attractive since the great "linguistic" potential was discovered in several highly social and intelligent species by means of intermediary artificial languages. Being applied to apes, dolphins, and grey parrots, this method has revealed astonishing mental skills in the subjects. However, surprisingly little is known as yet about natural communication systems even of those species that were involved in language-training experiments based on adopted human languages. Explorers of animal "languages" thus have met a complex problem of resolving the contradiction between their knowledge about significant "linguistic" and cognitive potential in some species and limitations in understanding their natural communications.

In this book, I attempt to give a multifaceted panorama of the methods developed for studying animal communication and of the most fascinating results in this field, as well as to highlight a conceptually distinct approach that is based on ideas of information theory. The main point of this approach is not to decipher animal signals, but to study a language and evaluate its capabilities by means of measuring the rate of information transmission. The experimental paradigm underlying this approach is to provide a situation in which animals have to transmit information quantitatively known to the experimentalist. The experiments on ants were elaborated and conducted together with the information theorist Boris Ryabko. The first results of this long-term work were reported at the International Information Theory Symposium in 1984, and then the obtained data were discussed at many conferences and published in biological and mathematical journals (see, e.g. Reznikova and Ryabko 1990, 1993, 1994, 2000, 2011; Ryabko and Reznikova 1996, 2009). This approach enabled us to discover a sophisticated symbolic "language" in highly social group–retrieving ant species. These insects were found to be able to transfer abstract information about remote events, to grasp regularities, and to use them to optimise their messages. These abilities distinguish ants among other "intellectuals", such as counting primates, crows and parrots, as one of the smartest groups of species.

This book starts with some basic notions and concepts used for studying animal language behaviour (Chap. 1) and the analysis of the main experimental approaches to studying intelligent communication in different species, from ants to whales (Chap. 2). The main features of social hymenopterans – ants and bees – as reliable agents for information transmission are considered in Chap. 3. In Chap. 4, language behaviour in highly social ant species is described. Details of their colony organisation underlying the process of information transmission are considered: team working and inter-relations between scouting ants and foragers. Chapters 5 and 6 make up the core of this book, being devoted to the use of ideas of information theory for studying ant "language" and the amazing intellectual skills of these animals, including their numerical competence.

I am grateful to Daniil Ryabko for his fruitful comments throughout the book and consultations on information theory.

This work was supported by Russian Science Foundation, grant 14-14-00603.

Novosibirsk, Russia Zhanna Reznikova

References

Reznikova Z, Ryabko B (1990) Information Theory approach to communication in ants. In: Gribakin FG, Wiese K, Popov AV (eds) Sensory systems and communication in arthropods. Birkhäuser Verlag, Basel, pp 305–307

Reznikova Z, Ryabko B (1993) Ants aptitude for the transmission of information on the number of objects. In: Wiese K, Gribakin FG, Popov AV, Renninger G (eds) Sensory systems of arthropods. Birkhäuser Verlag, Basel, pp 634–639

Reznikova Z, Ryabko B (1994) Experimental study of the ants communication system with the application of the Information Theory approach. Memorabilia Zoologica 48:219–236

Reznikova Zh, Ryabko B (2000) Using Information Theory approach to study the communication system and numerical competence in ants. In: Meyer JA, Berthoz A, Floreano D, Roitblat H, Wilson SW (eds) From animals to animats, vol. 6, Proceeding of the sixth international conference on simulation of adaptive behaviour. The MIT Press, Cambridge, MA, p 501–506

Reznikova Z, Ryabko B (2011) Numerical competence in animals, with an insight from ants. Behaviour 148(4):405

Ryabko B, Reznikova Z (1996) Using Shannon Entropy and Kolmogorov Complexity to study the communicative system and cognitive capacities in ants. Complexity 2:37–42

Ryabko B, Reznikova Z (2009) The use of ideas of information theory for studying "language" and intelligence in ants. Entropy 11(4):836–853

Contents

Communication, Speech and Language 1

1.1 Can Animals Exchange Meaningful Messages?

As early as in 1661 Samuel Pepys wrote in his diary about what he called a "baboon": "I do believe it already understands much English; and I am of the mind it might be taught to speak or make signs" (as cited in Wallmann 1992). More than 200 years later, Garner (1892) experimentally tried to learn the "monkey tongue." With the help of the phonograph, Garner conducted first playback experiments on several species of apes and monkeys in the Zoo. He artificially changed the elements of the animals' signals and then compared how the subjects reacted to variants of reproduced "words". In particular, Garner claimed to distinguish a word that meant "food" in the capuchin's "language" from others, which meant "bread" and "vegetables/fruits". Garner had thus anticipated some more recent findings.

Applying a similar experimental paradigm, Hauser and Marler (1992) found that rhesus macaques produce five acoustically distinctive vocalisations when they find food. Three vocalisations are restricted to the discovery of high quality, rare food items, whereas the other two are given while waiting for lower quality, common food items. It has been proven experimentally that the type of vocalisation produced is influenced by the type of food discovered and not by the discoverer's hunger level (Hauser 2000).

In the 1960s and 1970s, elegant but ambiguous experiments were conducted with highly socially intelligent animals, which were asked to pass some pieces of information to each other. In Menzel's (1973a, b) classic experiments, a group of chimpanzees living in an enclosure searched for hidden food. Menzel suggested that chimpanzees possess means for transferring information about both location and properties of objects, but it remained unclear how they did this. Evans and Bastian (1967) experimentally investigated the cooperative behaviour of dolphins, which could involve sophisticated communication. To get a fish, the pair of dolphins, separated by an opaque barrier, had to press the paddles in the correct order. The obtained results enabled researchers to suggest that dolphins can co-ordinate the behaviour of each other, probably by means of acoustic signals.

Zh. Reznikova, *Studying Animal Languages Without Translation: An Insight from Ants*, DOI 10.1007/978-3-319-44918-0_1

Despite these supportive experiments, the question of existence of developed "languages" in non-humans remained so far obscure. What can be called "language behaviour" in animals and how does it differ from communication and speech?

1.2 Communication

In animal studies the terms "communication" and "signalling" are used interchangeably. The term "communication" enjoys a wide variety of meanings, which is of no wonder because communication is a diverse and widespread phenomenon that serves a substrate of any social behaviour. It is difficult to imagine either social behaviour of any level lacking a means of communication or any transfer of information that had nothing to do with social relations. Most of the signals that animals send to one another communicate the intention, emotional state or identity of the sender. In the great majority of signalling interactions animals influence others rather than inform them. Dawkins and Krebs (1978) viewed communication as an inherently cooperative event, arguing that signalers were more likely to be manipulating than informing perceivers. On the contrary, Wilson (1975) and Maynard Smith and Harper (2003) qualify communication to include signals that convey information that perceivers benefit from receiving.

There are many definitions of the term "communication", and some of them include codes of specific signals which animals use to communicate with each other. For instance, Vauclair (1996) defines "communication" as follows:

"Communication consists of exchanges of information between a sender and a receiver using a code of specific signals that usually serve to meet common challenges (reproduction, feeding, protection) and, in group – living species, to promote cohesiveness of the group."

Definitions like this include both influence on the behaviour of members of the group, and behaviours specialized for transmitting encoded information. Recently Owren and co-authors (Owren et al. 2010) proposed a revised definition making influence rather than transmission of encoded information the central function of animal signalling. They defined animal signalling as "the use of specialized, species-typical morphology or behaviour to influence the current or future behaviour of another individual". This definition is quite inclusive and can accommodate all key phenomena currently considered to involve transmission of encoded information.

In this book the wide term "communication" is used that includes both unaware and unintentional sharing of information and language-like, symbolic communication. "Language behaviour" is considered separately, as the intentional transfer of information between members of a group.

In group-living animals communication, in its wide sense, serves many important functions such as (1) to advertise individual identity, presence, and behavioural predispositions; (2) to establish social hierarchies; (3) to synchronise the physiological states of a group during breeding seasons; (4) to monitor the environment collectively for dangers and opportunities; (5) to synchronise organised activities (migration, foraging).

Many species have well-developed forms of inter-individual exchange of information based on specific social signals related to courtship, defence of territories, ranking, care about offspring and other forms of social inter-relations. There is a huge body of examples concerning the use of expressive signals within a wide variety of species, from nematodes to elephants (see: Witzany 2014). In many taxa, signals involve multiple modes, such as visual, olfactory, auditory, tactile, and some others. For instance, lizards use contextual gestures such as submissive and aggressive circular forelimb waving, tail lashing, head bobbing, back arching, and so on (Sinervo and Lively 1996). Apes and humans use intentional gestures to communicate (Pollick and De Waal 2007).

To identify complex forms of animal communication that may be attributed to language behaviour, the important feature is purposiveness. It is often difficult to decide whether animals intend to share information with conspecifics or they use signalling inadvertedly. We can consider "language behaviour" the most complex form of animal communication that takes place when animals advisedly transfer the information to each other.

1.3 Speech

It is intuitively clear that communication it too broad a concept and speech is too narrow to consider as the form of information transfer in animals. Speech is a specially evolved system for language and serves as a very effective mean of communication. There are two major forms of language in humans: speech and the manual sign language of the deaf. Besides, many ways of signalling languages exist in human culture such as Morse code, drum signalling, flag signalling, whistles, semaphore, smoke signalling and so on.

In the great majority of definitions speech is stated to be a form of communication specific to humans. Phonetic assurance of speech in humans is supported by the coordinated actions of several articulators (jaw, lips, tongue, velum, and larynx) and by specific brain structures. The capacity to imitate vocalisations is confined to a few species of birds, certain marine mammals, certain monkeys and humans. However, perceptually, speech already has a unique status for the human infant within a few hours or days of birth. Newborns prefer speech to non-speech, and prefer their mother's voice to a stranger's. They can make most of the phonemic distinctions attested in the world's languages and they are able to distinguish languages they have never heard before on the basis of their rhythmical characteristics. Newborns are also able to detect the acoustic cues that signal word boundaries, discriminate words with different patterns of lexical stress and distinguish function words (e.g., *it*, *this*, *in*, *of*, *these*, *some*) from content words (e.g., *baby*, *table*, *eat*, *slowly*, *happy*) on the basis of their different acoustic characteristics (for a review see Gervain and Mehler 2010).

Developmental studies of infants from 1 to 6 months of age, based on the habituation-dishabituation experimental paradigm, have shown that infants can discriminate virtually any speech sound contrast on which they are tested, including contrasts not used in the surrounding languages. Over the second 6 months of life infants gradually cease to discriminate non-native sound contrast. During the period in which an infant is perhaps first attending to the different contexts in which words are used, it also gradually shifts speech sound contrasts that are functional in the surrounding language from those that are not. A real-world problem facing the human infant is how to segment the continuous acoustic stream of speech into functional units such as phonemes, words, and phrases. It has been shown that infants are equipped with mechanisms that enable them to extract abstract rules which subsequently may form the foundation upon which grammars are constructed (for reviews see: Hauser et al. 2002; Gervain and Mehler 2010).

Animal studies suggest that the computational abilities used to detect word boundaries and long-distance regularities may have evolved for purposes other than language and that they may be found in nonhuman animals. To reveal animals' capacities for detecting grammar-like regularities, Hauser and co-authors (Hauser et al. 2001) suggested the experimental paradigm based on artificial grammars. Series of experiments demonstrated that cotton–top tamarins (Fitch and Hauser 2004), rats (Toro and Trobalón 2005) and European starlings (Gentner et al. 2006) are able to segment synthesized speech streams and detect regularities associated with grammatical structures. It is important to note that, neuroanatomically, there is a direct connection from the cortical face area to the laryngeal motoneurons in humans but, say, not in monkeys. However, the principal neural prerequisites for decoding speech sounds seem to be already present in monkeys (Jürgens 2002; Weiss et al. 2002; Simonyan 2014).

As we have seen, speech is just one form of language, the most complex and likely to be confined to humans. Unlike the broad notion of communication, and the specific notion of speech, the concept of language should be more useful for reasoning about animal communication.

1.4 Language and "Intelligent Communication"

I suggest to distinguish between the concepts of "language" and "language behaviour" (or "Intelligent communication").

1.4.1 Language

The use of language is one of the most complex human skills and it is no wonder that many scientists define language in such a way that only human beings can be said to be capable of it. Chomsky (1959) suggests that the environment is mainly a trigger rather than a tutor for language acquisition, proposing that the ability to learn grammar is hard-wired into the brain ("mental grammar"). The same author (Chomsky 1986) considers language a prerogative of humans that is aimed to facilitate free expression of thought and clarify one's ideas, as well as to help establish social relations and to communicate information.

Behaviourally, language can be defined as a system of self-generated movements, composed of definable units, which can arbitrary represent some object, event, or intention on the part of the mover (Kimura 1979). Human language is the most sophisticated communicative system which includes thousands of spoken and written languages. As a reflective communication, language can be used to transfer ideas, to convey feelings, to lie and to exchange fantastic imaginations. It is a debated question whether any but human animals share these features. What can we interpret to be similar to human language depends on how we identify language and to what extent our concept of language allows comparisons between species.

Obviously, language cannot be described or defined by one single feature; it is a polymorphous concept. Hockett (1960, 1963) identified a range of characteristics that describes essential features of language. He compared selected properties of human language with properties of selected non-human communication systems, which he knew of from the ethological literature of his day such as bee dancing, stickleback courtship, herring gull care of offspring and gibbon calls. Modern data on animal communications have changed our imaginations about differences and similarities between human and non-human communication systems and intelligence (for a review see: Reznikova 2007). Hockett's system cannot acount for explanations of language evolution (Pinker and Jackendoff 2005; Wacewicz and Żywiczyński 2015). Besides, this system considers language a system of communication, whereas it is still debated whether language first evolved as a system of communication, or it evolved as a cognitive tool being specific to our species (Reboul 2015). However, Hockett's system of design features still has a high descriptive value. The following Hockett's list provides a useful framework for evaluating the "linguistic" skills of animals.

1. *Interchangeability.* Language is a two-way process that involves both sending and receiving the same set of signals. This is distinctive from some animal communications such as that of the stickleback fish. The stickleback fish make auditory signals based on gender (basically, the males say "I'm a boy" and the females say "I'm a girl"). However, a male fish cannot say "I'm a girl," although it can perceive such a signal. Thus, stickleback fish signals are not interchangeable. In many social species any member of a group can produce and perceive universal signals, so their communication systems meet this criterion.
2. *Specialization.* Language is not a by-product of some other biological function. Returning to the stickleback, the male reacts to biological aspects of females' messages such as swollen belly, and the female reacts to the change of the male's colouration, so the males' message is more specialised.
3. *Discreteness.* The basic units of language (such as sounds) can be organised into discrete units and classified as belonging to distinct categories. There is no gradual, continuous shading from one sound to another in the linguistics system, although there may be a continuum in the real physical world. Thus speakers will perceive a sound as either a [p] *or* a [b], but not as a blend, even if physically it falls somewhere between the two sounds. It is important that discrete units can be broken apart to form new signals.

4. *Arbitrariness of units*. Language is composed of discrete units, and the form of the signal does not depend on a referred thing. Messages consist of arbitrary units but not of icons describing an object.
5. *Displacement*. Language can be used to refer to things that are not present in space (here) or time (now). These requirements are known as "spatial displacement" and "temporal displacement" respectively. Human language allows speakers to talk about past and future. Speakers can also talk about things that are physically distant (such as other countries and other planets). They can even refer to things and events that do not actually exist. Many researchers consider that the honey bee's dance language meets this criterion. Bees are able to refer to objects that are distant in space and even in time. For instance, Lindauer (1960) reported that in the night bees performed dances signalling feeder sites that they had visited the previous day. Apes easily meet this criterion with the use of intermediary artificial language (see details in the Chap. 2) but we don't know about the potential of their natural communication.
6. *Semanticity*. Communication meets this criterion if signals (or words) convey specific meanings. We presume that for an animal a signal has meaning if it can activate a representation of the event to which it relates. Thus vervet monkeys appreciate the meanings of alarm calls they hear (Cheney and Seyfarth 1980, 1990), and it is likely that honey bees may appreciate the meaning of communications they receive (see details in the Chap. 2).
7. *The duality of patterning*. This is a property of language that allows a combinatorial structure on two levels: (i) phonology, where meaningless sounds called phonemes (i.e. the smallest meaning-differentiating elements of a language that do not themselves have meaning) can be combined into morphemes (i.e. the smallest meaningful elements) and words; and (ii) syntax, in which these morphemes and words can be combined into larger structures (see also: De Boer et al. 2012; more details in Chap. 2).
8. *Productivity*. Language can be used to produce an infinite variety of new messages from a limited vocabulary. Productivity is also called "creativity". In particular, language productivity enables the use of analogies, and by using of analogies human children adopt rules of grammar, or syntax. Productivity is a powerful property of language that makes it an open-ended system. The use of intermediary languages enables animals to meet this criterion. One challenge is to explore whether any natural animal communication system meets this requirement.
9. *Traditional (cultural) transmission*. Human language is not completely inborn. Although humans have a genetically-based capacity for language, they must learn, or acquire, their native language from other speakers. This is different from many animal communication systems in which the animal is born knowing their entire system. At the same time, as it will be described in the Chap. 2, cultural transmission plays a great role in the development of communication in some species of birds, monkeys, whales and may be some others species.

1.4.2 "Language Behaviour" and "Intelligent Communication"

The term "language behavior" usually refers to animal communication systems in which referential signals exist that can be compared with words in a human language. Some species use distinctive signals which seem to refer to definite external stimuli, for example, types of predators or kinds of food. If such signals provide receivers with sufficient information to determine the context underlying signal production which, in turn, allows them to predict environmental events, the signals are regarded as *functionally referential* (see more details in Chap. 2).

To avoid antropomorphic interpretations, we suggest to use the term "intelligent communication" (Reznikova 2007, 2012), which can be defined as a communication that includes referential signals and means for transferring information about remote events. I consider the terms "language behaviour" and "intelligent communication" interchangeably and use them this way further in the book.

One of the most striking examples of intelligent communication in non-human animals is the dance language of honey bees *Apis mellifera*. Karl von Frisch (1923, 1967) was the first who recognized how this unique form of communication allowed bees to share information about the location of food sources with nestmates: scouting bees memorise the location and features of the new food source that is kilometres away and transfer this knowledge to their nestmates in the hive by generating a specific encoded message that describes the direction and the distance from the hive to the goal. Von Frisch described the honey bee dance as "the most astounding example of non-primate communication that we know" (von Frisch 1967).

The use of some modern methods such as the robotic bee (Michelsen et al. 1992; Landgraf et al. 2011) and harmonic radar (Riley et al. 2005) has provided a quantitative description of how efficiently bees translate the code in the dance into flight to their destinations. However, important cognitive aspects of communication in bees still remain unclear (see details in Chap. 2).

Our studies have demonstrated that highly social ant species possess a system of intelligent communication that is possibly even more complex than that of honey bees (Reznikova and Ryabko 1994, 2000, 2011; Ryabko and Reznikova 1996, 2009). These results will be presented further in this book (Chaps. 3, 4, 5, and 6).

References

Bastian J (1967) The transmission of arbitrary environmental information between bottlenosed dolphins. In: Busnel RG (ed) Animal sonar systems – biology and bionics, vol. 2. NATO Advanced Study Institute/Lab. de Physiologie Acoustique, France, p 807–873

Cheney DL, Seyfarth RM (1980) Vocal recognition in free-ranging vervet monkeys. Anim Behav 28:362–367

Cheney DL, Seyfarth RM (1990) How monkeys see the world: inside the mind of another species. University of Chicago Press, Chicago

Chomsky N (1959) Verbal behavior by B. F. Skinner. Language 35(1):26–58

Chomsky N (1986) Knowledge of language: its nature, origins, and use. Praeger, New York

Dawkins R, Krebs JR (1978) Animal signals: information or manipulation. Behav Ecol Evol Approach 2:282–309
De Boer B, Sandler W, Kirby S (2012) New perspectives on duality of patterning: introduction to the special issue. Lang Cogn 4:251–259
Fitch WT, Hauser MD (2004) Computational constraints on syntactic processing in a nonhuman primate. Science 303(5656):377–380
Garner RL (1892) The speech of monkeys. New York: C. L. Webster and Co
Gentner TQ, Fenn KM, Margoliash D, Nusbaum HC (2006) Recursive syntactic pattern learning by songbirds. Nature 440(7088):1204–1207
Gervain J, Mehler J (2010) Speech perception and language acquisition in the first year of life. Annu Rev Psychol 61:191–218
Hauser MD (2000) A Primate dictionary? Decoding the function and meaning of another species' vocalizations. Cogn Sci 24:445–475
Hauser MD, Marler P (1992) How do and should studies of animal communication affect interpretations of child phonological development? In: Ferguson C, Menn L, Stoel-Gammon C (eds) Phonological development. York Press, Inc, Maryland, pp 663–680
Hauser MD, Newport EL, Aslin RN (2001) Segmentation of the speech stream in a non-human primate: statistical learning in cotton-top tamarins. Cognition 78(3):53–64
Hauser MD, Chomsky N, Fitch WT (2002) The faculty of language: what is it, who has it, and how did it evolve? Science 298:1569–1579
Hockett CD (1960) The origin of speech. Sci Am 203:89–96
Hockett CD (1963) The problem of universals in language. In: Greenberg JH (ed) Universals of language. The MIT Press, Cambridge, MA, pp 1–22
Jürgens U (2002) Neural pathways underlying vocal control. Neuroscience and Biobehavioral Reviews 26(2):235–258
Kimura D (1979) Neuromotor mechanisms in the evolution of human communication. In: Steklis HD, Raleigh MJ (eds) Neurobiology of social communication in primates. Academic Press, New York, pp 197–219
Landgraf T, Rojas R, Nguyen H, Kriegel F, Stettin K (2011) Analysis of the waggle dance motion of honeybees for the design of a biomimetic honeybee robot. PLoS One 6(8):e21354
Lindauer M (1960) Time compensated sun orientation in bees. Cold Spring Harbor Symposium on Quantitative Biology 25:372–377
Menzel EW (1973a) Chimpanzee spatial memory organization. Science 182:943–945
Menzel EW (1973b) Leadership and communication in young chimpanzees. In: Menzel EW (ed) Precultural primate behavior. S. Karger, Basel, pp 192–225
Michelsen A, Andersen BB, Storm J, Kirchner WH, Lindauer M (1992) How honeybees perceive communication dances, studied by means of a mechanical model. Behav Ecol Sociobiol 30(3–4):143–150
Owren MJ, Rendall D, Ryan MJ (2010) Redefining animal signaling: influence versus information in communication. Biol Philos 25(5):755–780
Pinker S, Jackendoff R (2005) The faculty of language: what's special about it? Cognition 95(2):201–236
Pollick AS, De Waal FB (2007) Ape gestures and language evolution. Proc Natl Acad Sci 104(19):8184–8189
Reboul AC (2015) Why language really is not a communication system: a cognitive view of language evolution. Front Psychol;6
Reznikova Z (2007) Animal intelligence: from individual to social cognition. Cambridge University Press, Cambridge
Reznikova Zh (2012) Intelligent communication in animals. In: Encyclopedia of the sciences of learning. Springer US, p 1597–1600
Reznikova Z, Ryabko B (1994) Experimental study of the ants communication system with the application of the Information Theory approach. Memorabilia Zoologica 48:219–236

Reznikova Zh, Ryabko B (2000) Using Information Theory approach to study the communication system and numerical competence in ants. In: Meyer JA, Berthoz A, Floreano D, Roitblat H, Wilson SW (eds) From animals to animats, vol. 6, Proceeding of the sixth international conference on simulation of adaptive behaviour. The MIT Press, Cambridge, MA, p 501–506

Reznikova Z, Ryabko B (2011) Numerical competence in animals, with an insight from ants. Behaviour 148(4):405

Riley JR, Greggers U, Smith AD, Reynolds DR, Menzel R (2005) The flight paths of honeybees recruited by the waggle dance. Nature 435(7039):205–207

Ryabko B, Reznikova Z (1996) Using Shannon Entropy and Kolmogorov Complexity to study the communicative system and cognitive capacities in ants. Complexity 2:37–42

Ryabko B, Reznikova Z (2009) The use of ideas of information theory for studying "language" and intelligence in ants. Entropy 11(4):836–853

Simonyan K (2014) The laryngeal motor cortex: its organization and connectivity. Curr Opin Neurobiol 28:15–21

Sinervo B, Lively CM (1996) The rock-paper-scissors game and the evolution of alternative male reproductive strategies. Nature 380:240–243

Smith JM, Harper D (2003) Animal signals. Oxford University Press, New York

Toro JM, Trobalón JB (2005) Statistical computations over a speech stream in a rodent. Percept Psychophys 67(5):867–875

Vauclair J (1996) Animal cognition: an introduction to modern comparative psychology. Harvard University Press, Cambridge, MA, pp 5–10

von Frisch K (1923) Über die Spräche der Bienen. Zoologische Iahrbucher-Abteilung für Allgemeine Zoologie und Physiologie der Tiere 40:1–119

von Frisch K (1967) The dance language and orientation of bees. Harvard University Press, Cambridge

Wacewicz S, Żywiczyński P (2015) Language evolution: why Hockett's design features are a non-starter. Biosemiotics 8(1):29–46

Wallmann JM (1992) Aping language (themes in the social sciences). Cambridge University Press, New York

Weiss D, Ghazanfar A, Miller CT, Hauser MD (2002) Specialized processing of primate facial and vocal expressions: evidence for cerebral asymmetries. In: Rogers L, Andrews R (eds) Cerebral vertebrate lateralization. Cambridge University Press, New York, pp 480–530

Wilson EO (1975) Sociobiology: the new synthesis. Harvard University, Cambridge, MA: Belknap Press

Witzany G, ed (2014) Biocommunication of Animals. Springer Science+Business Media Dordrecht

Do Animal "Languages" Need Translation? The Main Experimental Approaches to Studying Language Behaviour

2

The purpose of this chapter is to compare the three main experimental paradigms in animal language behaviour studies.

The first approach is aimed at direct decoding of animal signals. Although it is intuitively clear that many highly social species have to possess complex communications, only two types of natural messages have been decoded up to now. Namely, the symbolic honeybee dance language and acoustic signals of danger, which were decoded first for vervet monkeys and, later, for several other species.

The second approach is based on the use of intermediary artificial languages. Being applied to apes, dolphins, gray parrots and even dogs, this method has revealed astonishing mental skills in the subjects. It is important to note that this way to communicate with animals is based on adapted human languages.

The third, information-theoretic approach, which is the main subject of this book, is considered in details in Chaps. 5 and 6. The main point of this approach is not to decipher signals but to investigate the process of information transmission itself, by measuring the time duration that animals spend on transmitting messages of definite information content and complexity. The experimental paradigm is relatively simple. All we need to do is to "ask" subjects to transfer the intended quantity of information to each other. Specifically, in the experiments on ants, to organise the process of information transmission between animals a special maze was used, called the "binary tree" maze (Reznikova and Ryabko 1986, 1994), where the number and sequence of turns towards the goal correspond to the amount of information to be transferred. This experimental paradigm provides a conceptually distinctive way for studying important characteristics of natural communication system, such as the rate of information transmission, the complexity of transferred information and potential flexibility of communications.

Zh. Reznikova, *Studying Animal Languages Without Translation: An Insight from Ants*, DOI 10.1007/978-3-319-44918-0_2

2.1 Cracking Animals' Codes

"Deciphering" animal signals, that is, compiling species-specific "dictionaries" is the oldest and still the most popular approach to study communications in nature. Many researchers have tried to decode animal languages by looking for "letters", "words" and "phrases". With such an approach, it often remains unclear, which sounds and gestures have something to do with the "language" and which do not. In contrast to unequivocal mating signals, communications within groups and societies are often indistinct and difficult to follow.

Theberge and Pimlott (1969) noted that when studying wolves' ability to produce and distinguish subtle details of acoustic signals they were challenged by the problem of understanding a foreign culture's sounds lacking a relevant "dictionary" and any ideas about this culture. These authors suggested that wolves were able to transfer the information by changing certain units of their acoustic communication, but the only "word" they managed to decipher was a "sound of loneliness" that wolves produce being placed in isolation, anxious to join others.

It is likely that such simple and distinct signals, which always work uniformly and are thus amenable to decoding, can be found in different species. Bottlenose dolphins (*Tursiops truncatus*) develop individually distinctive signature whistles ("names") that they use to maintain group cohesion. Dolphins transmit identity information independent of the caller's voice or location (Janik et al. 2006; Kershenbaum et al. 2013). Recent studies on wolves and other canids showed that howls can encode individual and group identity (see review in: Kershenbaum et al. 2016). In honeybees the "buzz-run" ("Time to go!") signal is a ritualized form of the take-off behaviour, with the wing buzzing greatly exaggerated and other behavioural elements (running, butting and wiggling) added to increase the signal's detectability (Rittschof and Seeley 2008).

One can hypothesise that a bottleneck for decoding animals' signals is low repeatability of standard living situations, which could give keys for cracking animals' species-specific codes. In fact, animals sometimes behave similarly in repeatable situations, and if these repeatable behaviours are elicited by the distinctive repeatable signals, these behaviours can serve as keys for cracking animals' species-specific codes. The simple and distinct signals mentioned before for wolves, dolphins and bees are highly repeatable in very expressive and distinguishable situations and thus easily amenable to decoding.

Both in the honey-bee dance language and in acoustic signals of danger that were deciphered in several species expressive and distinctive signals correspond to repeatable and frequently occurring situations in the context of animals' life. However, in the field of studying animal "texts" there is a growing body of evidence about sophisticated structure of communications including symbolic meanings, semantics and syntax, as well as cultural "linguistic" traditions, all of which earlier have been attributed only to humans.

2.1.1 The Dance Language of Honeybees

The honey bee possesses the most complex animal natural "language" that has been decoded, at least partially. As it was mentioned in Chap. 1, successful forager honeybees are able to recruit other bees to a distant goal by specific "dance" movements together with other components of communications such as odours and sounds. von Frisch (1923, 1967) suggested that in the bees' dance language, an abstract, or symbolic, code is used to transmit the information about the direction and distance to a desired object. Years after von Frisch interpreted the symbolism of the dances other components of the bees' communication system have been discovered (for reviews, see: Seeley 1995; Michelsen 2003; Tautz 2008; I'Anson Price and Grüter 2015).

It is worth to note that honeybees can communicate by means of dance figures not only about food but also about the location of a new nest site (Lindauer 1961; Camazine et al. 1999; Tautz 2008), and that intellectual navigation requires cognitive resources (Menzel et al. 2005). Land and Seeley (2004) discovered a special "grooming invitation dance" by which bees "ask" their nestmates to help them to clean those parts of the body that they cannot reach themselves.

The suggestion that bees have a symbolic language was met with skepticism by many scientists, who favoured the idea that scent alone guides bees to a distant food source (Gould 1975). The experiments using an artificial model of a bee confirmed von Frisch's hypothesis: the dances do indeed represent a sophisticated form of communication. It is a natural idea to communicate with the bees trough mediation of a robotic bee. Indeed, a crucial test for any communication system is to modify it experimentally. The idea of using a model bee that dances and indicates the distance and direction of a target the other bees had never visited, was suggested by J.B.S. Haldane (1927) in his "The future of biology". Michelsen and colleagues (Michelsen 1993, 1999; Michelsen et al. 1990) built a bee of brass and beeswax with a piece of a razor blade to represent the wings. Their robot bee could buzz its wings at the requisite 280 cycles per second (Hz), waggle its bottom and dance in a circle. It could also deliver drops of sugar water to the dance watchers. Using a computer to control the robot, the researchers set their robotic bee to dancing, attempting to dissect the components of the dance. Dancing a normal waggle dance, this bee, like any robot, was a bit clumsy. Nevertheless, recruits in the hive used the information to fly to a target. On another day, researchers sent the bees in the opposite direction. Bees found the new targets regardless of the wind direction, correctly following messages from the robot bee that never left the hive itself. These studies allow us to reject the hypothesis that bees use the smell of the new food source as the main source of information in favour of the hypothesis of the use of dance parameters to decide where to fly.

2.1.2 Semantic Vocalisations in Animals: Words Without a Language

As it was mentioned before, while studying animal vocalisations, the question remains whether different call types are discrete or whether there are graded intermediates between call types. For instance, Barbary macaques, *Macaca sylvanus*, produce different variants of shrill barks in response to different predators, but there are no distinct boundaries between these call types (Hammerschmidt and Fischer 1998). Such graded variations have been found in a variety of mammal and birds species.

One of the main experimental methods for studying acoustic intelligent communication in animals is based on quantitative comparative analysis of signals and on play-back experiments where vocalisations recorded are played for animals either in their original form, or altered, for example, slowed down or with some elements added.

Lists of possibly meaningful signals have been compiled for several primate species. These signals include those for greetings, invitations for playing, sex, sharing food, predator alarm, intergroup treat, affiliation, infant distress, discontent, quarrel, nocturnal roll call, cooing with infants and so on (Lawick-Goodall 1968; Marler and Tenaza 1977; Snowdon 1986).

A representative "Vervet–English" dictionary has been first compiled by Struhsaker (1967). He found that vervet monkeys in the wild emit 25 discrete calls referring to different objects and situations. Nevertheless, the majority of signals were not distinct enough and not used with sufficient frequency. Only three "words" seemed to be understandable not only by monkeys but also by human observers. Three different sounds emitted by vervets for three different predators (leopards, eagles and snakes) resulted in three different reactions or escape responses. The calls appeared to function as representational, or semantic, signals. Seyfarth and Cheney (1990, 2003) applied playback experiments to reveal characteristics of vervets' communication. In particular, they addressed the question of whether animals can refer to the meanings of "words" rather than to their acoustic properties.

These findings have dramatically changed our understanding about acoustic communication in animals, which were previously believed to be only an expression of their current emotional status. Vervet communication is considered symbolic because the relationship between the signifier and the signified is arbitrary and purely conventional (Price et al. 2015). However, Seyfarth and Cheney (2003) accentuate limits of animal acoustic communication and its great difference with the sophisticated language of humans.

Since early vervet studies, human comprehension of animal "dictionaries" had been expanded to many other nonhuman species. For example, West African Diana monkeys, *Cercopithecus diana*, produce acoustically distinct alarm calls in reaction to leopards and crowned-hawk eagles, their main predators. Playback experiments have shown that monkeys treat these vocalisations as semantic signals, in the sense that they compare signals according to their meanings (Zuberbühler 2000). Prairie dogs were demonstrated to use different calls for four predators - human, hawk, coyote and domestic dog; they can even speed up or slow down their signals depending on whether the predator is running or walking through their colony (Slobodchikoff et al. 1991). Later it has been revealed that sciurid rodents (ground squirrels, prairie dogs and marmots) use different call types to communicate about different types of predators, to communicate different degrees of risk or response urgency, or to communicate both predator type and urgency (for a review see: Pollard and Blumstein 2012). Meerkats (*Suricata suricatta*) use alarm calls of one type in response to appearance of mammalian predators, primary jackals and African wild cats, alarm calls of another type in response to appearance of avian predators, primary the martial eagle, and alarm calls of a third type in response to snakes and to faecal, urine or hair samples of predators. As in many other species, meerkat pups possess inherited vocalisations but they definitely need to learn how to respond appropriately to the different alarm call types (Hollén and Manser 2007).

The alarm, food associated, and social calls of many species of primate, other mammals, and birds have been classified as functionally referential signals. As it was mentioned in Chap. 1, referential signals have been proposed to transmit not only characteristics about a sender's physical characteristics or motivational status, but also "information" about objects in the environment (Marler et al. 1992). By referring to external objects such as predators or food these signals elicit corresponding behavioural responses in the receivers as if the latter were responding directly to the external cue (Seyfarth et al. 1980; Evans and Evans 2007).

Food calls may be functionally referential, in the sense that they encode information about the kind, quantity and availability of food. Evidence that food calls provide sufficient information to evoke anticipatory feeding behaviour from conspecifics has been obtained using playback experiments on domestic chickens (Evans and Marler 1991). Ravens, *Corvus corax*, emit structurally discrete yells when they approach rich but defended food sources, and thus attract conspecifics (Heinrich 1999). Bugnyar et al. (2001) experimentally exposed a group of free-ranging ravens to six feeding situations. Researchers thus modelled distinct repeatable situations and examined what calls emitted by ravens correspond to which kind of food, of which quantity and availability. The different use of long and short yells relative to food availability suggests that short "whoo" calls provide information about the caller, whereas long "haa" calls may also provide information about the food itself. Further experiments have shown that raven food calls are individually distinct and that the birds may be capable of differentiating between food-calling individuals (Boeckle and Bugnyar 2012).

2.1.3 Syntax in the Wild

Syntax is defined as a set of rules for arranging words and phrases to impart meaning. It is a key feature of human language, as syntax allows a finite number of words, together with a few rules, to express an infinite number of concepts. In animal studies songs usually serve as examples of "syntax". Unlike calls, which are built of single syllables (sometimes repeated), songs include multiple syllables that are constructed in a specific order, often with repeated phrases. Songs are often hierarchically organized where notes are combined into syllables, syllables into motifs and motifs into phrases with multiple layers of repetition or periodicities (Marler and Slabbekoorn 2004). The use of rather complex hierarchically organised songs is well documented in birds (Slater 2003; Jarvis 2006), and in several mammals including gibbons (Mitani and Marler 1989; Clarke et al. 2006), whales (Payne and McVay 1971; Whitehead and Rendell 2014), bats (Bohn et al. 2009), hyraxes (Kershenbaum et al. 2012), wolves ("howling communication", see: Kershenbaum et al. 2016), and even domestic mice (Arriaga et al. 2012).

A more rigorous concept of syntax as applied to animal communication considers the ability to create meaningful combinations of signals. As described in Chap. 1, among Hockett's design features of language, there are two kinds of syntax: phonological and compositional. Humans have both, and until recently, non-human animals were believed to possess only compositional syntax (Marler 2000; Hurford 2011). A linguistically informed review of animal call sequences (Collier et al. 2014) argues that phonology in animal vocal systems is rare whereas compositional syntax is more widespread; however, there do exist some human languages possessing features without phonology, or lacking phonology altogether. In light of this, the authors hypothesize that syntax, present in all languages, evolved before phonology.

Here are several expressive examples of the use of compositional and phonological syntax in different taxa. Arnold and Zuberbühler (2006) revealed semantic combinations in male putty-nosed monkeys (*Cercopithecus nictitans*), which produce two acoustically distinct loud calls ("pyows" and "hacks") in response to a range of disturbances. Apart from these predator-specific call sequences, the authors found that males regularly combine the same two calls into a third structure, a "pyow–hack" (or P–H) sequence, which usually consists of one, two or three "pyows" followed by up to four hack. Play-back experiments and the use of GPS for tracking group movements showed that the monkeys' response to P–H sequences was not confined to the predator context, but was generally related to whether and how far the group moved.

The Japanese great tit (*Parus minor*) displays over ten different notes in its vocal repertoire and uses them either one by one or in combination with other notes. Experiments revealed that tits can "speak in phrases", that is, extract a new meaning from a combination of calls. Receivers extract different meanings from "ABC" (scan for danger) and "D" notes (approach the caller), and a compound meaning from "ABC–D" combinations. When hearing "ABC-D", receivers both scan for danger and fly to the caller. However, receivers rarely scan and approach when the order of the notes is artificially reversed ("D–ABC") (Suzuki et al. 2016).

Phonological syntax turns sounds that individually have no meaning into ones with meaning. Suffixes and prefixes in human languages are a good example, and other animals use strings of different notes that are never used individually. The Campbell's monkey adds an "oooh" sound to the end of its vocalizations to increase the intensity of the message, and this sound is never used on its own, that is, it serves as a suffix (Schlenker et al. 2014).

What about dolphins and wales, which seem do be good candidates for using syntax? According to Whitehead and Rendell (2014), wild dolphins and whales do not use syntax, or use it only in an exceedingly simple way, although they can learn and use some syntax when taught by humans (see the next issue of this chapter). As the authors give this, "when we first heard the patterned codas of sperm whales, an immediate hypothesis was syntax. Maybe the "4+2" coda meant "go to [4], my calf ([+2])". But the more we studied codas and the way whales used them, the more it became apparent that these strange patterns of clicks were important in bonding and other relationships between individual whales and groups of whales, and the less likely it was that they were conveying complex information syntactically." (page 289).

2.1.4 Intelligence Communication and Experience

The role of experience in the development of communication in young social non-human and human animals is a matter of special interest to ethologists, linguists and psychologists. The earliest experiment on this topic with human beings that survives comes in the form of a legend told by Herodotus: the Egyptian Pharaoh Psammetichus I ordered two infants to be brought up in a remote place by a shepherd who was forbidden to speak in their presence. After 2 years the children began to speak, and the word that they repeated most often was *becos*, which turned out to be the Phrygian word for bread. Psammetichus concluded that the capacity for speech is innate, and that the natural language of human beings is Phrygian.

Similar experiments with non-human animals have started relatively recently. The research has been mainly concentrated on vocal communications. The concept of vocal learning is rather complex (see: Egnor and Hauser 2004) and includes: (i) vocal comprehension learning, which occurs when the appropriate response to a vocalization is learned; (ii) vocal production learning, which arises when specific spectrotemporal features of a vocalization depend on specific auditory experiences; (iii) vocal usage learning, which arises when the social or ecological context in which the call is used is learned. We know now that a few mammalian species and some birds can modify their vocalizations in response to auditory experience.

Experimental approaches to study vocal production learning are similar with Psammetichus', that is, naïve individuals were reared in isolation or in the presence of another species (cross-fostering experiments). Both experimental paradigms have been successfully applied to birds (see details below), and recently to dolphins. In the study of Favaro et al. (2016) a female Risso's dolphin (*Grampus griseus*) was stranded at an early age and was subsequently raised in a group of bottlenose dolphins (*Tursiops truncatus*). This cross-fostered animal produced vocal signals more akin to those of its Tursiops poolmates than those of Risso's dolphins in the wild. This finding suggests that social experience is a major factor in the development of the vocal repertoire in Risso's dolphins.

Currently the songbird is the most successful model system for vocal production learning. Thorpe (1958) reared chaffinches *Fringilla coelebs* in isolation and determined which aspects of their songs are innate and which have to be learned from adults. This study has been greatly expanded upon by Marler (1997) and many others (for reviews see: Slater 2003; Jarvis 2006). Vocal production learning in birds displays many significant parallels to humans. For example, both groups require auditory experience with species-typical vocalizations during an early sensitive period in order to develop normal adult vocal behaviour, and both exhibit greater vocal plasticity during development than in adulthood, although some songbirds retain vocal plasticity throughout life (Doupe and Kuhl 1999). However, unlike human language acquisition, vocal production learning in songbirds is largely restricted to song–a vocalisation that in most species is produced only by males. Paradoxically, closely related to our species non-human primates lack the behavioural and neural elements classically associated with vocal learning (Janik and Slater 1997), although they can make small changes to innately specified vocalisations (Egnor and Hauser 2004).

Based on the evidence available, we can say that vocal learning, although very rare, has a surprisingly discontinuous distribution among animals, from whales and elephants to small rodents. For instance, a recent study has demonstrated that mice require auditory feedback in order to maintain their ultra-sound songs produced by males in the mating context, although these songs are not quite as sophisticated or advanced as in humans or song-learning birds (Arriaga et al. 2012). The vocal repertoire of African and Asian elephants, with about 8–10 distinct call types, exhibits great vocal plasticity, context-dependent flexibility and elements of vocal learning (Poole et al. 2005; Stoeger and Manger 2014).

2.1.5 Do Ants Need "Babbling"?

Do nonhuman animals need "babbling", like human infants? And why speculate about "babbling" in ants? Although we know that ants can exchange meaningful messages by means of antennal movements (for a review see: Reznikova 2007), almost nothing is known about the ontogenetic development of antennal communication in them.

There are still only a few examples of babbling behaviour in animals. All of them concern vocal practice, and not gestural communication, although it would be interesting to compare ants' early antennal movements with first awkward communication gestures in young primates and elephants. First, there are striking parallels between the way human infants learn to speak and birds learn to sing: in both cases, learning is the product of the interaction of predispositions and specific experience (Wilbrecht and Nottebohm 2003). Vocal imitation in human infants and in some orders of birds relies on auditory-guided motor learning during a sensitive period of development. It proceeds from "babbling" (in humans) and "subsong" (in birds) through distinct phases towards the full-fledged communication system. Language development and birdsong learning have parallels at the behavioural, neural and genetic levels (Bolhuis et al. 2010).

Infant babbling behaviour is also known in a few non-human primates (Elowson et al. 1998). For example, in the pygmy marmoset (*Cebuella pygmaea*) infants produce long repetitions of mixed call types that are similar to the vocalizations of adult conspecifics and resemble the babbling bouts produced by human infants (Snowdon and Elowson 2001). The evidence of babbling in nonprimate mammals has been found in the sac-winged bat (*Saccopteryx bilineata*), a species with an unusually large vocal repertoire (Knörnschild et al. 2006). In this species pups produce renditions of all known adult vocalization types during bouts of vocalizations. In contrast to humans (Goldstein et al. 2003) and monkeys (Snowdon and Elowson 2001) in which babbling is associated with increased social interactions with group members, babbling in infant sac-winged bats does not meet any social response, and serves for training rather than for communication with conspecifics (Knörnschild et al. 2006).

Even in *Myrmica* ants one can detect a hint at acoustic babbling in the study of pupae stridulation (Casacci et al. 2013): The sounds generated by worker pupae were similar to those of workers but were emitted as single pulses rather than in the long sequences characteristic of adults.

The mode of life of social insects is wholly based on communication that includes different sensory modalities such as acoustic, visual, odour and tactile signals, (Hölldobler and Wilson 1990; Tautz 2008). There is evidence, though scarce, of learning aspects of communication in bees and ants. Experiments with mixed colonies of two species of honeybee indicated that Asiatic honeybees can decode the dances of European honeybees, and this suggests the possibility of social learning between the two honeybee species (Su et al. 2008). In ants, social learning between a pair of species was experimentally demonstrated (Reznikova 1982; see details in Chap. 3). Long-term laboratory experiments revealed a symbolic communication system in highly social ant species which is possibly even more complex than that of honey bees (Reznikova and Ryabko 1994, 2011; see details in Chap. 5). The cognitive flexibility of ants' communications (Reznikova 2008) enables us to hypothesise the existence of a learning component in their "language". However, it is not known yet whether young ants need lessons from adults to develop their "language" capacities.

Antennal mode of communication is a good candidate for studying the role of social experience in the development of communication in ants. As early as in 1899, Wasmann suggested the existence of "tactile (or antennal) code" as the way that ants use to communicate, and this hypothesis has been developed by many researchers (for a review see Reznikova 2008). Measuring exactly the duration of antennal contacts between scouting ants and their followers we revealed the complex symbolic "language" in red wood ants (Reznikova and Ryabko 1994, 2011; see details in Chaps. 5 and 6).

Ontogenetic development of antennal communication in ants had been studied in *Camponotus vagus*. Filming of pairwise interaction among callow workers and older ones during trophallaxis (liquid food transfers from mouth to mouth) revealed that units of antennal movements become more complex and coordinated in maturing ants (Bonavita – Cougourdan and Morel 1984; Morel 1986). However, these investigations were limited to trophallaxis and did not consider ants' communication in a context of other vital situations. As far as I know, after these studies there were no attempts to study the role of social experience in ontogenetic development of antennal communication in ants.

We have launched the first – and yet preliminary - study on the development of communication in *Myrmica rubra* ants (Atsarkina et al. in press) studying their behaviour and communication at the early stage of the imaginal ontogenesis. First, let us consider the main peculiarities of this early period of imaginal lifetime in ants.

Ants are known to begin learning some signals and even communicating long before eclosion. Cross-fostering experiments, in which larvae were introduced in an alien adoptive colony, where they developed until the pupal stage and then transferred back to their original colony, revealed preimaginal olfactory learning in formicine ants (Isingrini et al. 1985) and in *Aphaenogaster senilis* (Signorotti et al. 2016). In *M. scabrinodis* sclerotised pupae generate sounds by a stridulatory organ and thus attract attention of workers (Casacci et al. 2013). Thus, newly hatched ants are equipped with some skills of olfactory and acoustic communication, and they are ready to gain further experience in different forms of communication including "antennal code".

The early imaginal period has been described in some ant studies as a sensitive period for a variety of behavioural patterns and tasks such as cocoon recognition (Jaisson 1972; Le Moli and Mori 1982), recognition of nestmates (le Moli and Mori 1984), aggression (Errard 1984; Reznikova and Iakovlev 2008), and brood care (Champalbert and Lachaud 1990). Early social experience has a strong effect on the development of trophallaxis in *Camponotus vagus* since callow workers deprived of relations with mature workers from before hatching showed a slower behavioural ontogenesis than workers that hatched within their colony (Morel 1986).

Our with colleagues experiments on *Formica* suggested that, like many vertebrate species, ants have innate templates for recognition of some vitally important objects, and they need some triggers to start particular behavioural patterns, and some experience to develop these behaviours. For example, *Formica* s.str. have an innate template for recognising enemies such as carabid beetles (Reznikova and Dorosheva 2004, 2013), and ladybird imagines (Novgorodova 2015). Deprivation experiments with red wood ants (Reznikova and Novgorodova 1998; Reznikova 2007) showed that to develop such a sophisticated behavioural stereotype as aphid tending, ants needed trigger stimuli from their symbionts. When the mode of interaction with aphids had been completed, naïve ants still needed much more social experience with older workers to develop optimal division of labour within groups of aphid tenders.

In our study on the development of antennal communication in ants (Atsarkina et al. in press) a polygynous *Myrmica* colony was housed on the laboratory arena, and we selected pairs of callow (1–7 days) and adult ants for a detailed observation of their interactions. Ants were placed in pairs into two linked Petri dishes on a lighted arena. We then analysed video records of contacts in pairs "adult – adult", "adult – callow" and "callow – callow". We compared frequences and duration of different elements in their behaviour repertoire, and revealed significant differences in the behaviour of adult and callow workers, as well as in their mode of communication.

In the presence of callows adults decrease their mobility and prefer staying near the young, frequently giving them long antennal contacts. Adults initiate both

Fig. 2.1 (**a**) An adult worker actively touches the callow worker's body whereas a callow ant stands still keeping its antennae rigid, wide open (Photo by Natalia Atsarkina). (**b**) Callow ants "talk" with each other (Photo by Natalia Atsarkina). (**c**) Callow ants display awkward bouts of slow antennal movements (Photo by Natalia Atsarkina)

antennae-antennae and antennae - body contacts with young ants in most of the cases. An adult worker actively touches the callow worker's body whereas a callow ant stands still keeping its antennae rigid, wide open (Fig. 2.1a). On the contrary, during encounters among two callow ants there are no differences between frequencies and durations neither of long and short contacts, nor of contacts starting from antennal and body touching. In general, callow ants are more "talkative" than adult ones. They display larger duration of contacts, and contacts constitute the larger part of the whole time budget in callow ants than in adult ones. It is worth noting that an average duration of contacts between two callow workers is the same as between callows and adults. This means that callow workers are equally ready to "talk" both with each other and with adult ones (Fig. 2.1b).

That adult workers prefers to stay near callow ones contacting with it enables us to suggest that young ants possess attractiveness for adults. One can propose some behavioural (and rather anthropomorphic) hypotheses like "babyness", "loneliness" in the artificial world outside a nest, and so on. As far as "babyness" is concerned, callow ants are known to possess some character distinctive features, strongly associated with age. Callow workers seem to be characterized by a "cuticular chemical insignificance" followed by a "chemical integration" period when they acquire the gestalt of the colony and learn the associated template (Lenoir et al. 1999).

Antennal movements are slow and clumsy in young ants, and they often switch from communication to other activities (Fig. 2.1c). It is likely that patterns of antennal movements in callows change gradually. We suggest that callow ants need awkward bouts of slow antennal movements which somehow resemble the babbling vocal bouts produced by human infants. This is a very distant parallel; however, we consider it promising for future research in the field of development of communication in social animals. It is likely that young ants need prolonged "babbling" contacts with adult ants to gain the experience of communication. Further experiments are needed to reveal whether ants need "babbling" to develop their antennal communications and to establish their social network. In particular, it would be interesting to reveal whether young ants can transfer information about a target by their clumsy and slow antennal movements.

2.1.6 Cultural "Linguistic Traditions" in Animals

"Linguistic traditions" in animals are often connected with their ability to maintain geographically distinctive "dialects" based on vocal learning within groups. "Dialects" are well known in birds (Podos and Warren 2007), and they have also been revealed in prairie dogs (Slobodchikoff et al. 1991) chipmunks (Gannon and Lawlor 1989), marmosets (de la Torre and Snowdon 2009), bats (Esser and Schubert 1998), hyraxes (Kershenbaum et al. 2012), wolves (Kershenbaum et al. 2016), and honey bees (Su et al. 2008).

The most striking example of the role of social learning in the development of vocal traditions is the "cultural evolution" in dolphins and whales (for reviews see: Janik 2014; Whitehead and Rendell 2014). Cultural traditions in cetaceans include learning both single signals and sequences of them. For example, the individually distinctive signature whistle ("name") in bottlenose dolphins (mentioned before in this section) is influenced by vocal learning; each individual develops its own novel, unique frequency modulation pattern after listening to whistles in its environment (Fripp et al. 2005). One of the species with relatively well-studied vocal culture is the killer whale *Orcinus orca* (see reviews in: Filatova et al. 2012; Filatova et al. 2015; Fig. 2.2). Resident killer whales live in stable matrilineal units that may group together, forming subpods and pods. Pods have distinct dialects comprising a mix of unique and shared call types; calves adopt the call repertoire of their matriline through social learning. Like in birds, though a fine structure of stereotyped calls in killer whales is learned, the range of signals that can be learned is limited by genetically inherited predispositions that can be more or less strict depending on the selective advantages of some cultural traits over others (Lachlan and Feldman 2003).

2.2 Direct Dialogs with Animals by Means of Intermediary Languages

The use of intermediary languages for studying "linguistic" and intellectual potential of animals has changed the general perception of animal communication and intelligence. About 40 years ago, it could have been difficult to imagine that animals

Fig. 2.2 A killer whale *Orcinus orca* pursuing researchers in order to give them an interview (one can see a green cord of a hydrophone within a boat) (Courtesy of T. Ivkovich, the FEROP Project)

can learn to associate arbitrary signs with meanings, to generate new symbols with new meanings, and to use these signs to communicate simple statements, requests and questions; to refer to objects and events displaced in time and space; to classify novel objects into appropriate semantic categories; and to transmit their knowledge to peers and offspring.

There are many excellent books and reviews written by researchers who carried out projects on teaching sign languages to apes (Patterson and Linden 1981; Savage-Rumbaugh and Lewin 1994; Fields et al. 2007), dolphins (Herman 2010; Herzing and Johnson 2015), and an African grey parrot (Pepperberg 2008). In this Section, we briefly consider how this method has influenced the development of animal language behaviour studies.

As early as 1929, Yerkes and Yerkes (1929) suggested the idea of teaching chimps the sign language used by deaf people, in order to make the dialog between humans and non-humans possible. Ulanova (1950) possibly was the first who taught a primate to use gesture signs to refer to desired things. She taught a rhesus macaque to make finger gestures to obtain rewards of different kinds. Different gestures corresponded to different food items (nuts, pieces of apples, bread, strawberry and garden radish) and drinks (coffee, tea and milk).

Several projects aimed at trying to teach apes a human language started in 1966, from the project of Gardner and Gardner (1969) who began teaching a chimpanzee the use of American Sign Language (ASL). Similar studies using ASL were carried out using other chimpanzees (Terrace 1979; Fouts 1997), gorillas (Patterson 1978) and an orangutan (Miles 1993). Alternative approaches to teaching apes a language have been elaborated basing on special devices. Premack (1971) elaborated an

Fig. 2.3 Kanzi the bonobo with E. Sue Savage – Rumbaugh using lexigrams on portable keyboard to communicate (Copyright © 2006 Great Ape Trust; courtesy of Duane Rumbaugh)

artificial language in which the words were plastic figures, which varied in shape, size, texture and colour. Plastic words were arbitrary. Sarah, Premack's brightest chimpanzee, was taught nouns, verbs, adjectives, pronouns and quantifiers; she was also taught to use the words for same–different, as well as negation and compound sentences. Premack's technique has been developed further to a project supervised by Rumbaugh (1977), in which the symbols serving as words—or "lexigrams" as they were called—were displayed on a keyboard connected to a computer.

One of the most interesting studies of language learning in primates is that of the bonobo Kanzi (and later of several other bonobos) with the use of lexigram keyboard (Rumbaugh and Savage-Rumbaugh 1994; Fig. 2.3). Series of tests showed that bonobo could both understand and produce sentences. The researchers claim that Kanzi's communications, made with lexigrams and gestures, were structured according to rules, so he used a simple version of grammar (Savage-Rumbaugh et al. 1998; Lyn et al. 2011).

The African Grey parrot (*Psittacus erithacus*) Alex had been taught by Pepperberg (Pepperberg 1999; Pepperberg 2008) to use English as an intermediary language. Alex was trained first to speak using names of objects (e.g. "key", "grain", "paper" and so on). Whenever Alex named an object correctly, he was praised and then allowed to eat the object or to play with it. Being trained in this manner during many years, Alex was able to use more than 100 English words correctly to refer to all objects in his laboratory environment that play a role in his life. He answers correctly an astonishing number of questions regarding these objects, such as "What object is blue?", "What shape is wood?", "How many are wool?".

Herman and colleagues studied dolphins' linguistic skills focusing on their language comprehension rather than on language production (Herman 1980, 2010; Herman et al. 1999). The researchers concentrated on dolphins' receptive competencies, mainly on their capabilities of processing both semantic and syntactic information. The primary syntactic devise used in the studies was word order. Dolphins were shown to be capable of understanding that word order changes meaning. Dolphins' ability to understand whether things are present or missing gave a possibility to test whether they are capable of symbolic reference. Herman and Forestell (1985) constructed a syntactic frame consisting of an object name followed by a gestural sign called "Question". For example, the two-item gestural sequence called "basket question" asks whether a basket is present in the dolphin's habitat. The dolphin could respond "yes" by pressing a paddle to his right or "no" by pressing a paddle to his left. Over a series of such questions, with the particular objects present being changed over blocks of trials, the dolphin was able to report accurately whether the named object was absent or present. These results indicate that the gestures assigned to objects were understood referentially by the dolphin, i.e. that the gestures acted as symbolic references to those objects. Results from similar studies on sea lions, which revealed much the same findings as with dolphins, can be found in (Schusterman and Krieger 1986; Schusterman et al. 2002).

In general, dialogs with several highly social and intelligent species by means of intermediary languages have revealed astonishing "linguistic potential" in these species, which includes such important properties of a developed language as propensity for categorisation, creativity and displacement. However, this method does not allow us to judge about the complexity of animals' natural communications.

The information-theoretic approach is designed to study quantitative characteristics of natural communicative systems, which, in turn, can be used to reason about certain properties of animal intelligence. This approach is considered in Chaps. 5 and 6.

References

Arnold K, Zuberbühler K (2006) The alarm-calling system of adult male putty-nosed monkeys, *Cercopithecus nictitans martini*. Anim Behav 72(3):643–653

Arriaga G, Zhou EP, Jarvis ED (2012) Of mice, birds, and men: the mouse ultrasonic song system has some features similar to humans and song-learning birds. PLoS One 7(10):e46610

Atsarkina N, Panteleeva S, Reznikova Zh. *Myrmica* ants are talkative when young: do they need babbling? (in press)

Boeckle M, Bugnyar T (2012) Long-term memory for affiliates in ravens. Curr Biol 22(9):801–806

Bohn KM, Schmidt-French B, Schwartz C, Smotherman M, Pollak GD (2009) Versatility and stereotypy of free-tailed bat songs. PLoS One 4(8):e6746

Bolhuis JJ, Okanoya K, Scharff C (2010) Twitter evolution: converging mechanisms in birdsong and human speech. Nat Rev Neurosci 11(11):747–759

Bonavita – Cougourdan A, Morel L (1984) Les activites antennaires au cours des contacts trophallactique chez la Fourmi *Camponotus vagus* Scop. Ont-elles valeur de signal? Insect Soc 31:113–131

Bugnyar T, Kijne M, Kotrschal K (2001) Food calling in ravens: are yells referential signals? Anim Behav 61(5):949–958

Camazine S, Visscher PK, Finley J, Vetter RS (1999) House-hunting by honey bee swarms: collective decisions and individual behaviors. Insect Soc 46(4):348–360

Casacci LP, Thomas JA, Sala M, Treanor D, Bonelli S, Balletto E, Schönrogge K (2013) Ant pupae employ acoustics to communicate social status in their colony's hierarchy. Curr Biol 23(4):323–327

Champalbert A, Lachaud JP (1990) Existence of a sensitive period during the ontogenesis of social behaviour in a primitive ant. Anim Behav 39(5):850–859

Clarke E, Reichard UH, Zuberbühler K (2006) The syntax and meaning of wild gibbon songs. PLoS One 1(1):e73

Collier K, Bickel B, van Schaik CP, Manser MB, Townsend SW (2014) Language evolution: syntax before phonology?. In: Proceedings of the Royal Society B (vol. 281, No. 1788). The Royal Society, London, p 20140263. http://dx.doi.org/10.1098/rspb.2014.0263

de la Torre S, Snowdon CT (2009) Dialects in pygmy marmosets? Population variation in call structure. Am J Primatol 71(4):333–342

Doupe AJ, Kuhl PK (1999) Birdsong and human speech: common themes and mechanisms. Annu Rev Neurosci 22(1):567–631

Egnor SR, Hauser MD (2004) A paradox in the evolution of primate vocal learning. Trends Neurosci 27(11):649–654

Elowson AM, Snowdon CT, Lazaro-Perea C (1998) Babbling' and social context in infant monkeys: parallels to human infants. Trends Cogn Sci 2(1):31–37

Errard C (1984) Evolution, en fonction de l'âge, des relations sociales dans les colonies mixtes hétérospécifiques chez les fourmis des genres *Camponotus* et *Pseudomyrmex*. Insect Soc 31(2):185–198

Esser KH, Schubert J (1998) Vocal dialects in the lesser spear-nosed bat *Phyllostomus discolor*. Naturwissenschaften 85(7):347–349

Evans CS, Evans L (2007) Representational signalling in birds. Biol Lett 3(1):8–11

Evans CS, Marler P (1991) On the use of video images as social stimuli in birds: audience effects on alarm calling. Anim Behav 41(1):17–26

Favaro L, Neves S, Furlati S, Pessani D, Martin V, Janik VM (2016) Evidence suggests vocal production learning in a cross-fostered Risso's dolphin (*Grampus griseus*). Anim Cogn:1–7

Fields WM, Segerdahl P, Savage-Rumbaugh S (2007) The material Practices of ape language research. In: The Cambridge handbook of sociocultural psychology, Cambridge University Press, NY, p 164–202

Filatova OA, Ford JK, Matkin CO, Barrett-Lennard LG, Burdin AM, Hoyt E (2012) Ultrasonic whistles of killer whales (*Orcinus orca*) recorded in the North Pacific (L). J Acoust Soc Am 132(6):3618–3621

Filatova OA, Samarra FI, Deecke VB, Ford JK, Miller PJ, Yurk H (2015) Cultural evolution of killer whale calls: background, mechanisms and consequences. Behaviour 152(15):2001–2038

Fouts R (1997) Next of Kin, my conversations with chimpanzees. William Morrow and Company, Inc., New York

Fripp D, Owen C, Quintana-Rizzo E, Shapiro A, Buckstaff K, Jankowski K, Wells R, Tyack P (2005) Bottlenose dolphin (*Tursiops truncatus*) calves appear to model their signature whistles on the signature whistles of community members. Anim Cogn 8(1):17–26

Gannon WL, Lawlor TE (1989) Variation of the chip vocalization of three species of Townsend chipmunks (genus *Eutamias*). J Mammal 70(4):740–753

Gardner RA, Gardner BT (1969) Teaching sign language to a chimpanzee. Science 165:664–672

Goldstein MH, King AP, West MJ (2003) Social interaction shapes babbling: testing parallels between birdsong and speech. Proc Natl Acad Sci 100(13):8030–8035

Gould JL (1975) Honey bee recruitment: the dance language controversy. Science 189:685–693

Haldane JBS (1927) The future of biology. Possible Worlds:139–153

Hammerschmidt K, Fischer J (1998) Maternal discrimination of offspring vocalizations in Barbary macaques (*Macaca sylvanus*). Primates 39(2):231–236

Heinrich B (1999) The mind of the raven. Harper Collins, New York

Herman LM (1980) Cognitive characteristics of dolphins. In: Herman LM (ed) Cetacean behavior: mechanisms and functions. Wiley Interscience, New York, pp 363–429

Herman LM, Forestell PH (1985) Reporting presence or absence of named objects by a language-trained dolphin. Neuroscience and Biobehavioral Reviews 9(4):667–691

Herman LM (2010) What laboratory research has told us about dolphin cognition. Int J Comp Psychol 23(3):310–330

Herman LM, Abichandani SL, Elhajj AN, Herman EYK, Sanchez JL, Pack AA (1999) Dolphins (*Tursiops truncatus*) comprehend the referential character of the human pointing gesture. J Comp Psychol 113:1–18

Herzing DL, Johnson CM (2015) Dolphin communication and cognition: past, present, and future. MIT Press, Cambridge, MA

Hölldobler B, Wilson EO (1990) The ants. Harvard University Press, Cambridge, MA

Hollén LI, Manser MB (2007) Motivation before meaning: motivational information encoded in meerkat alarm calls develops earlier than referential information. Am Nat 169(6):758–767

Hurford JR (2011) The origins of grammar: language in the light of evolution II, vol 2. Oxford University Press, Oxford, New York

I'Anson Price R, Grüter C (2015) Why, when and where did honey bee dance communication evolve? Front Ecol Evol 3:125

Isingrini M, Lenoir A, Jaisson P (1985) Preimaginal learning as a basis of colony brood recognition in the ant *Cataglyphis cursor*. Proc Natl Acad Sci USA 82:8545–8547. doi:10.1073/pnas.82.24.8545

Jaisson P (1972) Preliminary note on the ontogenesis of nest care behavior in young brown ants (*Formica polyctena* Först): role of a probable imprinting mechanism. Comptes rendus hebdomadaires des séances de l'Académie des sciences Série D: Sciences naturelles 275(23):2721

Janik VM (2014) Cetacean vocal learning and communication. Curr Opin Neurobiol 28:60–65

Janik VM, Slater PJ (1997) Vocal learning in mammals. Adv Stud Behav 26:59–100

Janik VM, Sayigh LS, Wells RS (2006) Signature whistle shape conveys identity information to bottlenose dolphins. Proc Natl Acad Sci 103(21):8293–8297

Jarvis ED (2006) Selection for and against vocal learning in birds and mammals. Ornithol Sci 5(1):5–14

Kershenbaum A, Ilany A, Blaustein L, Geffen E (2012) Syntactic structure and geographical dialects in the songs of male rock hyraxes. Proc R Soc Lond B Biol Sci 279(1740):2974–2981

Kershenbaum A, Sayigh LS, Janik VM (2013) The encoding of individual identity in dolphin signature whistles: how much information is needed? PLoS One 8(10):e77671

Kershenbaum A, Root-Gutteridge H, Habib B, Koler-Matznick J, Mitchell B, Palacios V, Waller S (2016) Disentangling canid howls across multiple species and subspecies: Structure in a complex communication channel. Behav Processes 124:149 157

Knörnschild M, Behr O, von Helversen O (2006) Babbling behavior in the sac-winged bat (*Saccopteryx bilineata*). Naturwissenschaften 93(9):451–454

Lachlan RF, Feldman MW (2003) Evolution of cultural communication systems: the coevolution of cultural signals and genes encoding learning preferences. J Evol Biol 16(6):1084–1095

Land BB, Seeley TD (2004) The grooming invitation dance of the honey bee. Ethology 110(1):1–10

Lawick-Goodall J (1968) The behaviour of free-living chimpanzees in the Gombe Stream Reserve. Anim Behav Monogr 1:161–311

Le Moli F, Mori A (1982) Early learning and cocoon nursing behaviour in the red wood-ant Formica lugubris Zett. (Hymenoptera: Formicidae). Ital J Zool 49(1–2):93–97

le Moli F, Mori A (1984) The effect of early experience on the development of "aggressive" behaviour in Formica lugubris Zett. (Hymenoptera: Formicidae). Z Tierpsychol 65:241–249

Lenoir A, Fresneau D, Errard C, Hefetz A (1999) Individuality and colonial identity in ants: the emergence of the social representation concept. In: Information processing in social insects. Birkhäuser, Basel, p 219 237
Lindauer M (1961) Communication among social bees. Harvard University Press, Cambridge
Lyn H, Greenfield PM, Savage-Rumbaugh S, Gillespie-Lynch K, Hopkins WD (2011) Nonhuman primates do declare! A comparison of declarative symbol and gesture use in two children, two bonobos, and a chimpanzee. Lang Commun 31(1):63–74
Marler P (1997) Three models of song learning: evidence from behavior. J Neurobiol 33(5):501–516
Marler P (2000) Origins of music and speech: insights from animals. In: The origins of music. Cambridge, Mass, MIT Press, p 31–48
Marler PR, Slabbekoorn H (2004) Nature's music: the science of birdsong. Academic Press, Amsterdam, Boston
Marler P, Tenaza R (1977) Signaling behavior of apes with special reference to vocalization. In: Sebeok T (ed) How animals communicate. Indiana University Press, Bloomington, pp 965–1033
Marler P, Evans CS, Hauser MD (1992) Animal signals: motivational, referential, or both. In: Papousek H, Jürgens U, Papousek M (eds.) Nonverbal vocal communication: comparative and developmental approaches. Cambridge University Press, Cambridge, p 66–86
Menzel R, Greggers U, Smith A, Berger S, Brandt R, Brunke S, Bundrock G, Hülse S, Plümpe T, Schaupp F, Schüttler E, Stach S, Stindt J, Stollhoff N, Watzl S (2005) Honey bees navigate according to a map-like spatial memory. Proc Natl Acad Sci 102:3040–3045
Michelsen A (1993) The transfer of information in the dance language of honeybees: progress and problems. J Comp Physiol A Sens Neural Behav Physiol 173:135–141
Michelsen A (1999) The dance language of honeybees: recent findings and problems. In: Hauser MD, Konishi M (eds) The design of animal communication. The MIT Press, Cambridge, MA, pp 111–113
Michelsen A (2003) Signals and flexibility in the dance communication of honeybees. J Comp Physiol A 189(3):165–174
Michelsen A, Andersen BB, Kirchner W, Lindauer M (1990) Transfer of information during honeybee dances, studied by means of a mechanical model. In: Bribakin FG, Wiese K, Popov AV (eds) Sensory systems and communication in arthropods. Birkhäuser Verlag, Basel, pp 284–300
Miles HL (1993) Language and the orangutan: the old "person" of the forest. In: The great ape project. St. Martin's Griffin, New York, p 42–57
Mitani JC, Marler P (1989) A phonological analysis of male gibbon singing behavior. Behaviour 109(1):20–45
Morel L (1986) Ontogenesis of the antennal activity associated with food transfer in the callow worker ant. Dev Psychobiol 19:413–426
Novgorodova TA (2015) Role of social and individual experience in interaction of the meadow ant *Formica pratensis* (Hymenoptera: Formicidae) with ladybird imagines and hoverfly larvae. Insect Sci 22(3):440–450
Patterson FG (1978) The gestures of a gorilla: language acquisition in another pongid. Brain Lang 5:72–97
Patterson FG, Linden E (1981) The education of Koko. Holt Rinehart & Winston, New York
Payne RS, McVay S (1971) Songs of humpback whales. Science 173:585–597.l
Pepperberg IM (1999) The Alex studies. Harvard University Press, Cambridge, MA
Pepperberg IM (2008) Alex & Me: how a scientist and a parrot uncovered a hidden world of animal intelligence–and formed a deep bond in the process. HarperCollins, NY
Podos J, Warren PS (2007) The evolution of geographic variation in birdsong. Adv Study Behav 37:403–458
Pollard KA, Blumstein DT (2012) Evolving communicative complexity: insights from rodents and beyond. Phil Trans R Soc B 367(1597):1869–1878

Poole JH, Tyack PL, Stoeger-Horwath AS, Watwood S (2005) Animal behaviour: elephants are capable of vocal learning. Nature 434(7032):455–456

Premack D (1971) Language in chimpanzee? Science 172:808–822

Price T, Wadewitz P, Cheney D, Seyfarth R, Hammerschmidt K, Fischer J (2015) Vervets revisited: a quantitative analysis of alarm call structure and context specificity. Sci Rep 5:1–11

Reznikova Z (2007) Animal intelligence: from individual to social cognition. Cambridge University Press, Cambridge

Reznikova Z (2008) Experimental paradigms for studying cognition and communication in ants (Hymenoptera: Formicidae). Myrmecol News 11:201–214

Reznikova J(Zh.) (1982) Interspecific communication among ants. Behaviour 80;84–95

Reznikova Z, Dorosheva H (2004) Impacts of red wood ants *Formica polyctena* on the spatial distribution and behavioural patterns of ground beetles (Carabidae). Pedobiologia 48(1):15–21

Reznikova Z, Dorosheva E (2013) Catalog learning: carabid beetles learn to manipulate with innate coherent behavioral patterns. Evol Psychol 11(3):147470491301100304

Reznikova Zh, Iakovlev IK (2008) Development of aggression as a possible basis of "professional" specialization in ants. In: Doklady biological sciences (vol. 418, No. 1). MAIK Nauka/Interperiodica, Moscow, p 56–58

Reznikova Z, Novgorodova TA (1998) The importance of individual and social experience for interaction between ants and symbiotic aphids. Dokl Biol Sci 359:173–175

Reznikova Z, Ryabko B (1986) Investigations of ant language by methods of Information Theory. Probl Inf Transm 21:103–108

Reznikova Z, Ryabko B (1994) Experimental study of the ants communication system with the application of the Information Theory approach. Memorabilia Zoologica 48:219–236

Reznikova Z, Ryabko B (2011) Numerical competence in animals, with an insight from ants. Behaviour 148(4):405

Rittschof CC, Seeley TD (2008) The buzz-run: how honeybees signal 'Time to go!'. Anim Behav 75(1):189–197

Rumbaugh DM (1977) Language learning by a chimpanzee. The Lana project. Academic Press, New York

Rumbaugh DM, Savage-Rumbaugh ES (1994) Language in comparative perspective. In: Mackintosh NJ (ed) Animal learning and cognition. Academic Press, New York, pp 307–333

Savage-Rumbaugh ES, Lewin R (1994) Kanzi: the ape at the brink of the human mind. John Wiley & Sons, New York

Savage-Rumbaugh ES, Shanker SG, Taylor TJ (1998) Apes, language and the human mind. Oxford University Press, Oxford

Schlenker P, Chemla E, Arnold K, Lemasson A, Quattara K, Keenan S, Stephan C, Ryder R, Zuberbühler K (2014) Monkey semantics: two 'dialects' of Campbell's monkey alarm calls. Linguist Philos 37(6):439–501

Schusterman RJ, Krieger K (1986) Artificial language comprehension and size transposition by a California sea lion (*Zalophus californianus*). J Comp Psychol 100:348–355

Schusterman RJ, Kastak CR, Kastak D (2002) The cognitive sea lion: meaning and memory in the laboratory and in nature. In: The cognitive animal. Empirical and theoretical perspectives on animal cognition. The MIT Press, Cambridge, MA, p 217–228

Seeley TD (1995) The wisdom of the hive: the social physiology of honey bee colonies. Harvard University Press, Cambridge, MA

Seyfarth RM, Cheney DL (1990) The assessment by vervet monkeys of their own and another species' alarm calls. Anim Behav 40:754–764

Seyfarth RM, Cheney DL (2003) Signalers and receivers in animal communication. Annu Rev Psychol 54:145–173

Seyfarth RM, Cheney DL, Marler P (1980) Vervet monkey alarm calls semantic communication in a free-ranging primate. Anim Behav 28:1070–1094

Signorotti L, Jaisson P, d'Ettorre P (2016) Larval memory affects adult nest-mate recognition in the ant *Aphaenogaster senilis*. Proc R Soc B 281:20132579. doi:10.1098/rspb.2013.2579
Slater PJ (2003) Fifty years of bird song research: a case study in animal behaviour. Anim Behav 65(4):633–639
Slobodchikoff CN, Kiriazis J, Fischer C, Creef E (1991) Semantic information distinguishing individual predators in the alarm calls of Gunnison's prairie dogs. Anim Behav 42(5):713–719
Snowdon CT (1986) Vocal communication. In: Mitchell G, Erwin J (eds) Comparative primate biology, vol. 2A: behaviour, conservation and ecology. Alan R. Liss, New York, p 495–530
Snowdon CT, Elowson AM (2001) 'Babbling' in pygmy marmosets: development after infancy. Behaviour 138(10):1235–1248
Stoeger AS, Manger P (2014) Vocal learning in elephants: neural bases and adaptive context. Curr Opin Neurobiol 28:101–107
Struhsaker T (1967) Behavior of vervet monkeys (Cercopithecus aethiops). University of California Press, Berkeley
Su S, Cai F, Si A, Zhang S, Tautz J, Chen S (2008) East learns from West: asiatic honeybees can understand dance language of European honeybees. PLoS One 3(6):e2365
Suzuki TN, Wheatcroft D, Griesser M (2016) Experimental evidence for compositional syntax in bird calls. Nat Commun;1–7. doi:10.1038/ncomms10986
Tautz J (2008) The buzz about bees: biology of a superorganism. Springer Science & Business Media. Heidelberg, Springer
Terrace HS (1979) Nim: a chimpanzee who learned Sign language. Knopf, New York
Theberge JB, Pimlott DH (1969) Observations of wolves at a rendezvous site in Algonquin Park. Can Field Nat 83:122–128
Thorpe WH (1958) The learning of song patterns by birds, with especial reference to the song of the chaffinch *Fringilla coelebs*. Ibis 100(4):535–570
Ulanova LI (1950) Shaping notations expressing need for food in monkeys. In: Protopopov VP (ed) Russian. In Studying of high nervous activity by means of natural experiments. Gosmedizdat, Kiev, p 103–114
von Frisch K (1923) Über die Spräche der Bienen. Zoologische Iahrbucher-Abteilung für Allgemeine Zoologie und Physiologie der Tiere 40:1–119
von Frisch K (1967) The dance language and orientation of bees. Harvard University Press, Cambridge
Whitehead H, Rendell L (2014) The cultural lives of whales and dolphins. University of Chicago Press, Chicago
Wilbrecht L, Nottebohm F (2003) Vocal learning in birds and humans. Ment Retard Dev Disabil Res Rev 9(3):135–148
Yerkes RM, Yerkes AW (1929) The great apes: a study of anthropoid life. Yale University Press, New Haven
Zuberbühler K (2000) Referential labelling in Diana monkeys. Anim Behav 59(5):917–927

3 Social Hymenopterans as Reliable Agents for Information Transmission

The majority of models consider cognitive skills and individual interactions in social insects redundant, and assume that their behaviour is based on what is called "swarm intelligence," that is, it is only governed by collective decision making (Miller et al. 2013). These models are used to describe reactions of large groups as a whole to relatively simple stimuli from the environment. However, certain highly social animals, and, in particular, some species of social insects, exhibit the phenomenon of cognitive specialisation, that is, the ability of some individuals to learn faster within specific domains. This means that "cognitive specialists" form associations between some stimuli easier than between other stimuli and thus learn more readily certain behaviours (Reznikova 2007a). The presence of "cognitive specialists" facilitates the tuning of integrative reactions of the whole animal community to unpredictable influences in its changing environment (Reznikova 2012b).

In this book I concentrate on individual decision making in social insects and their abilities to transfer exact information among colony members by means of *distant homing*. Distant homing here means that messages about remote events come from a scouting individual, without other cues such as a scent trail or direct guiding (for reviews see: Reznikova 2007a, 2008). The most striking example of distant homing is the honeybee dance language (see details in Chap. 2) and the symbolic "language" of highly social ant species (see details in Chaps. 4, 5 and 6). Massive amounts of experimental evidence support the conclusion that distant homing is based on individual cognitive abilities. In ants distant homing is closely connected with the levels of social organisation of their colonies. In this chapter we consider the features of social Hymenopterans (mainly ants) as agents of sophisticated forms of information transmission, namely, their intelligence, levels of social organisation, modes of communication and peculiarities of the colony structure which precede distant homing.

Zh. Reznikova, *Studying Animal Languages Without Translation: An Insight from Ants*, DOI 10.1007/978-3-319-44918-0_3

3.1 Individual Cognitive Skills in Social Hymenopterans

It has been possible to apply the concept of intelligence to hymenopterans when Schneirla (1929) and Thorpe (1950) experimentally demonstrated that ants (two species of *Formica*) and solitary wasps perform almost as well as rats and dogs in maze learning and detour tasks. In our efforts to comprehend the multifaceted panorama of animal intelligence and cognition, social hymenopterans can serve as a touchstone. These insects are known to combine highly integrative colony organisation with sophisticated cognitive skills implemented by individual tiny brains.

Bees, wasps and ants possess several specific mechanisms, which should help them in the efficient perception and learning (for reviews see: Reznikova 2007b; Srinivasan 2010; Avarguès-Weber and Giurfa 2013). Mazokhin-Porshnyakov (1969; Mazokhin-Porshnyakov and Kartsev 2000) first experimentally demonstrated that honey bees and social wasps are capable of abstraction, extrapolation and solving rather sophisticated discrimination tasks at a level of dogs and monkeys. In particular, several individually trained bees were able to distinguish between chains consisted of paired and unpaired small figures and thus capable of concept formation concerning "twoness" (for details see: Reznikova 2007a). The early experiments of Reznikova (1979a, 1983) demonstrated that red wood are able to grasp a particular regularity of the discrete movements of a feeder and thus "plan" their actions. In the laboratory experiment red wood ants were presented with a set–up shaped as a "fan" consisting of 11 paper strips 15 cm in length, with 15° angle between them. During each session, ants were let to visit the set–up every 10 min, and each time a drop of syrup was placed on the subsequent strip, moving clockwise: first in sector 1, then in sector 2, and so on, as shown on Fig. 3.1. The feeder thus moved in discrete steps, on each step increasing the angle from its initial position, from the first to the eleventh strip. In total, 16 sessions were conducted with 2-day intervals between them. The "fan" was situated vertically during 11 sessions and horizontally during 5 sessions. Ants succeed in "planning" their visits correctly after the 4th session. They visited each time the strip subsequent to the one they had visited the previous time. They did not hesitate to apply their experience obtained with the vertical "fan" to the horizontal one.

Feats of intelligence in bees have been demonstrated in many experimental studies by means of different traditional approaches for studying animal intelligence, such as sequence learning (Collett et al. 1993; Kartsev 2014), maze learning (Zhang et al. 1999), the so-called Delayed Match to Sample (DMTS) task (Zhang et al. 2005), and even the Delayed Symbolic Match to Sample (DSMTS) task. In the last case the experimental subject had to use the identity of a sample stimulus (which can be A or B) to choose between two other comparison stimuli (C and D) that were presented simultaneously in a subsequent chamber. The subject had to learn to choose stimulus C if the sample is A, and stimulus D if the sample is B. In other words, the subject had to learn to associate A with C, and B with D. In this task, neither of the comparison stimuli matched the sample in the literal sense: the matching was symbolic. Until recently, only vertebrates—humans, monkeys, dolphins, crows and pigeons—were known to be able to learn DSMTS tasks (for reviews see:

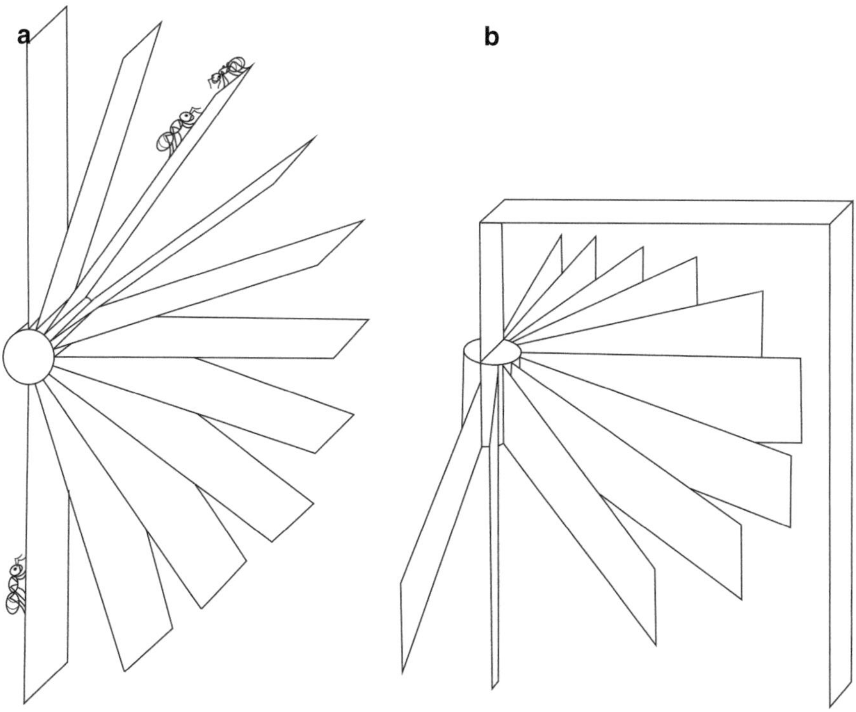

Fig. 3.1 A set–up shaped as a "fan" consisting of 11 paper strips. The feeder moved in discrete steps, on each step increasing the angle from its initial position, from the first to the eleventh strip (**a**) the vertical "fan", (**b**) the horizontal one. (Adapted from Reznikova 1983)

Reznikova 2007a; Smirnova et al. 2015). However, Zhang et al. (1999) revealed this ability in honey bees using a multilevel maze with two stages of decision chambers and with a sample stimulus presented at the entrance.

Bees were found to be able to form the concepts of symmetry and asymmetry. Bees also appeared to be particularly good at distinguishing between patterns that are topologically distinct, such as a disc versus a ring. Honeybees rapidly learn conceptual relationships involving visual stimuli (Giurfa et al. 1999). Concepts such as "same," "different," "above/below of" or "left/right" are also well mastered by bees (Avarguès-Weber and Giurfa 2013). Bees displayed the ability to recognize and discriminate between images of human faces and can evidently commit faces to long-term memory. Similarly to humans, bees discriminate familiar faces poorly when the images are presented upside down (Dyer et al. 2005).

There is much evidence that ants of different species are good at maze learning (Schneirla 1929; Evans 1932; Vowles 1965; Reznikova 1982; Karas and Udalova 1999; Cammaerts Tricot 2012), and at complex route learning (Rosengren and Fortelius 1986; Czaczkes et al. 2013), as well as in visual discrimination of shapes (Cammaerts and Cammaerts 2015). Red wood ants display the level of cognitive skills that has been known before only in vertebrate animals (Reznikova 2007a,

2012b). Ants demonstrated the capacity for rule extraction, that is, they can grasp regularities in the "texts" consisting of two letter-alphabet "R"(right torn) and "L" (left turn). Ants use these regularities in order to optimise their messages: "3 LR" instead of "LRLRLR". Moreover, these insects possess numerical competence, that is, they can use numbers of "branches" in order to pass the message about which one contains food, and even use simple arithmetic (Reznikova and Ryabko 1994, 2003, 2011). These cognitive abilities are described in details in Chaps. 5 and 6.

All this enables us to suggest that social hymenopterans can do many clever things and their ability to think abstractly is closely connected with their intelligent communication. Which forms of communication ants use is related to the levels of social organisation and their foraging strategies. These characteristics of ants' social life are considered in Sect. 3.2.

3.2 Recruitment Strategies and Levels of Social Organisation in Ants

Life-styles in ant species are highly varied. More than 12,000 species of ants have been described, and they display a great variety of specific adaptations at different levels of sociality (Hölldobler and Wilson 1990; Heinze 2008). Besides the primary division of labour between the reproductive caste and the worker caste, there exists a further division of labour among workers, whereby each member of the colony specialises in a subset of tasks required for successful group functioning (Reznikova 2007a, 2011). Different tasks have different levels of cognitive demand (see details in Sect. 3.3). For instance, scouting requires more cognitive resources than, say, defending a nest entrance. Even within the frame of the same task different ranges of complexity are required. In ants and bees scouts navigate rapidly and accurately to food sources and relevant sites for housing relying on their individual exploration, learning and memory.

Ants employ a range of recruitment strategies based on different communication methods for directing nestmates to foraging sites (For reviews see: Hölldobler and Wilson 1990; Detrain et al. 1999; Jackson and Ratnieks 2006; Leonhardt et al. 2016). Communication is useful when food resources are found that are larger than can be exploited by a single forager. Recruitment strategies in ants can be classified into the following main categories: (1) tandem running, where a successful forager leads a recruit, (2) mass recruitment, by which recruiters returning from a food source to the nest lay a chemical trail that guides their nestmates to the source, (3) group recruitment where the scout first lays a chemical trail upon return to the nest and subsequently leads a small group of recruits along this trail to the source, and (4) group recruitment based on distant homing without laying odour trails.

Most ant species use only a single foraging strategy, but some species use several, and they can also switch between different strategies depending on external stimuli coming from the environment, such as the size of the available food sources, as well as internal stimuli, such as colony growth.

Here are a few out of many examples concerning ants' ability to adjust their foraging strategies to changing resources. *Formica cunicularia* ants switch from sole foraging to mass recruitment depending on the size of the insect prey available (Reznikova 1975). Similar results have been obtained by Cerdá et al.(2009) on the gypsy ant *Aphaenogaster senilis*. Several species of harvester ants of the genus *Messor* are known as using odour trail when the seeds are aggregated and foraging individually when the seeds are dispersed over a large area (Hahn and Maschwitz 1985). The strategy of predation in several species of Ponerinae also depends on the prey encountered, and they use different strategies such as individual foraging, tandem running, mass recruitment, cooperative and group recruitment (Hölldobler and Traniello 1980; Schatz et al. 1997). The tree-dwelling African ant *Polyrhachis laboriosa* uses individual foraging when the food resource is small, whereas for large permanent food sources scouts use group recruitment. The choice of the foraging strategy is determined by the first forager, which modifies its behaviour according to the volume of the food supply (Mercier and Lenoir 1999).

The colony growth is closely connected with levels of social organisation in ants, and serves as an important factor determining the choice of foraging strategies. Levels of social organisation in ants vary between species but still little is known about the scale of diversity of social systems within species.

For example, in *Formica* species the lower level is the use of a feeding territory by a colony inhabiting a single domicile, and on the highest level is the super-colony. A super-colony in ants is an association of interconnected colonies inhabiting single nests. This system is closely connected with polygyny in this group of species: only polygynous colonies can form supercolonies (Fortelius et al. 1993). One of the largest examples of super-colonies in *Formica* is an enormous polydomous population of the Japanese ant *F. yessensis*, which covered an area of 2.7 km^2 and contained an estimated 306 millions workers and one million queens (Higashi and Yamauchi 1979). A supercolony of *F. lugubris* in Switzerland consisted of about 1200 interconnected nests occupying an area of 70 ha (Cherix 1980). A supercolony of polygyneous families of *F. pratensis* with similar forager densities was found in the vicinity of the southern borders of the geographic area of this species in the valley of Ili river in South Kazakhstan (Reznikova 1983). In the central parts of its area in steppe and forest-steppe zones *F. pratensis* inhabits single nests, and each monogyneous colony protects its own territory (Reznikova 1979b).

With such a great range of levels of social organisation *F. pratensis* can serve as a good candidate for elucidating the question whether a single colony can switch between several levels of social organisation. It is of particular interest in this situation that this species uses a rather peculiar organisation of foraging (Reznikova 1979b). Inhabiting a single domicile, a colony uses trench-like foraging routes. Some ants go along these routes towards their individual foraging plots on the foraging territory, whereas others circulate along the segments of the route waiting for being mobilised by scouts that search for food sources on the territory. These workers circulating along the routes constitute the "candidate pool" for being called by scouts.

In the field experiment the ants' dynamic density on one of the foraging routes of a single colony was artificially increased with the use of long barriers, which initially forced ants to concentrate within a relatively small territory. In response, ants set up auxiliary stations along the route, and they even accepted several fertilised mating queens in their colony. The colony thus changed its status to the "temporal polygyny". After that, the ants readily switched their recruitment mode in accordance with their new territorial organisation. Scouts stopped mobilising ants from the route. Instead, upon finding food on foraging plots, scouts attracted workers roaming in the vicinity by making alerting movements. This mode of recruitment, called "kinopsis" (Reznikova 1982) or "directional recruitment" is rather typical for supercolonies of *Formica* s.str. (Rosengren and Fortelius 1987). Amazingly, this reconfiguration took only a few weeks.

Similar patterns, although within narrow limits, have since been revealed experimentally in *F. cunicularia* by Reznikova (1983) and Novgorodova (2003). It is very likely that some ant species possess "toolkits" of social systems, and it would be interesting to investigate how their individual cognitive skills change in accordance with social systems. It is possible that dormant cognitive abilities would become observable when an ant colony switches to a higher level of social organisation. It is natural to expect that ant species with different colony design develop communication systems of different levels of complexity.

3.3 Distribution of Cognitive Responsibilities Within Ant Colonies

In this Section we consider only three out of many existing modes of distribution of cognitive responsibilities within an ant colony with different recruitment strategies, as representative and clearly distinct examples.

3.3.1 Mass Recruitment

In many ant species, patterns of interactions among individuals are governed by rules of self-organisation, and fairly simple units generate complicated behaviour by the group as a whole. This large group of species include mass foraging ants that broadcast guidance information potentially to all foragers, in the form of a trail network marked with varying amounts and types of pheromone. For example, Pharaoh's ant, *Monomorium pharaonis* uses a short-lived (approximately 20 min) attractive trail pheromone to guide foragers to the rewarding feeding sites, and a long-lived pheromone to maintain the trail network which can be explored during several days. The negative pheromone (repellent) is placed locally in the network, immediately after trail bifurcations on a non-rewarding branch, thus playing the role of "no entry" signal (Jackson and Ratnieks 2006).

In colonies of mass foraging ant species individuals essentially act as replaceable parts. Dornhaus et al. (2008) marked individually all members of a laboratory colony of *Temnothorax albipennis* in order to examine how ants make decisions of task allocation and to quantify worker efficiency. She measured work completed per time, in four different tasks: honey and protein foraging, collection of nest-building material, and brood transports during colony migration. Workers appeared to differ in their ability to perform different tasks. Surprisingly, this variation is not utilised by the colony – worker allocation to tasks is unrelated to their ability to perform them. One can hypothesise that members of mass-foraging species, though display variable capacities that could precede "professional specialisation," cannot develop this system in their colonies. There is no difference in cognitive responsibilities within families of mass-foraging ant species.

3.3.2 Sole Foraging

It has been demonstrated by many authors that members of solely foraging ant species are more agile and clever than members of mass-recruiting and even group-retrieving species. In sole-foraging species each forager searches for food by itself and thus undertakes long trips and makes its own decisions whether to catch a prey and to take it home or not.

For example, members of the genus *Cataglyphis* search for food individually at a distance over tens of meters. Like honey bees, these ants rely mainly on visual cues when negotiating their way through open or cluttered environments. In my early field experiments with *C. aenescens* in the desert of Kazakhstan, I offered a red wing of Orthoptera to an ant, and when it readily seized this "red flag" it was easy to observe how far it ran straight to the nest entrance; the longest way was up to 40 m (Reznikova 1983). In their field experiments with several species of *Cataglyphis*, Wehner and colleagues revealed intriguing details of orientation, learning and memory in these ants (for reviews see: Wehner 2003, 2009). The navigational toolkit of *Cataglyphis* includes a skylight (polarization) compass, a "path integrator", in which the animal keeps a continuously updated record of its current direction and distance from some reference point as it moves away from that place, and view-dependent methods for recognizing places and following landmark routes. The same ant can learn different routes for the outward and homeward journeys and for journeys leading to different feeding sites.

It is worth to note that *Cataglyphis* ants have only about 6 days to perform feats of intelligent orientation. For example, in the North African desert ant *Cataglyphis bicolor* an age-related polyethism divides the colony into interior workers up to an average age of 28 days followed by outdoor foraging period (the final in their lives) of about 6 days (Wehner 1972; Schmid-Hempel 1984). Their complex visually guided orientation is governed to a great extent by the high degree of neuronal plasticity in visual input regions of their mashroom bodies (Stieb et al. 2010).

Solely foraging ant species display high cognitive abilities. A striking example of the difference between solely foraging and group retrieving ant species in their "intellectual potential" came from the field study of ants' ability to navigate "round mazes" (Reznikova 1975, 1982, 2007a). Ants were presented with mazes containing

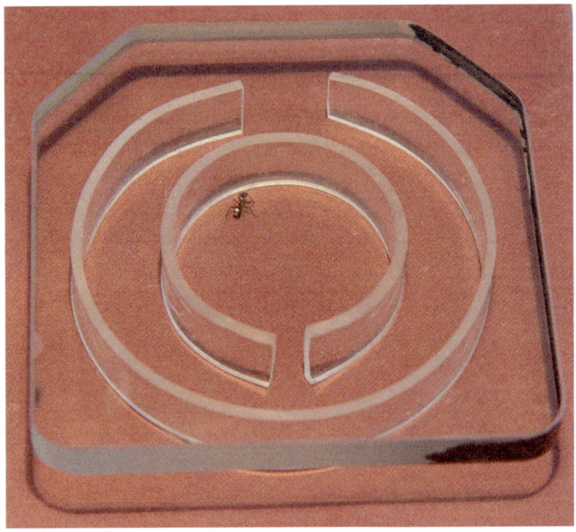

Fig. 3.2 A round maze with three *circles* (Photo by Ivan Iakovlev)

food pellets. The maze of the first level of complexity contained one circle with one entrance. This could hardly be called a "maze" because all the ants had to to was to enter the round box and get the food. The maze of the second level consisted of two concentric circles placed one inside the other in such a way that their entrances were placed on the opposite sides. The most complex maze consisted of three circles (Fig. 3.2a, b). In different series of field experiments, 100 mazes of the same level of complexity were placed in the ants' feeding territory. Two groups of ant species were compared, one group with a sole-foraging system (*F. cunicularia*, subgenus *Serviformica*) and other with group retrieving (*F. pratensis*, subgenus *Formica* s.str.). It turned out that in sole foraging species nearly all active foragers could successfully obtain food pellets from the most complex maze. In group – foraging *F. pratensis* not all of the ants coped with the maze problem. Instead, there were only "top ten" individuals which successfully navigated mazes consisting of two circles, that is, about 10 % of foragers.

What is of special interest here is that the solely foraging species serves as a scout for a group-retrieving one. In the system of interspecies hierarchy firstly experimentally discovered by Reznikova (1980, 1999), *F. pratensis* and *F. cunicularia* play the roles of the dominant and the subdominant correspondingly. In the Siberian steppes *F. pratensis* is a territorial dominant over *F. cunicularia*, driving their workers away from favorite nest sites and food finds. Occasionally, *F. cunicularia* foragers can steal food items that have been temporarily laid on the ground by *F. pratensis* foragers, but in general they have to rely on quickness and luck to gather food before *F. pratensis* arrive on the scene. For their part, *F. pratensis* play a role of scroungers, using *F. cunicularia* as scouts. Reznikova (1982, 2007a) tested this form of interspecies transfer of information in the field experiments presenting the

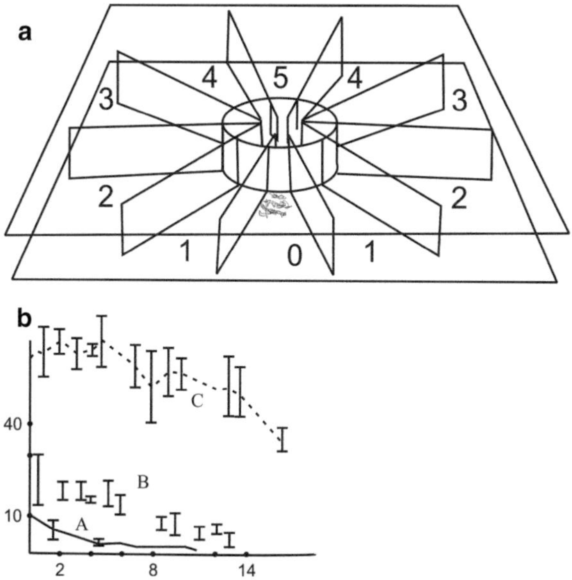

Fig. 3.3 Experiments on interspecies transfer of information in ants. (**a**) Circular maze with the baits in one of the sectors; 0–5 – numbers of the sectors. (**b**) The change in the number of errors during the training: *A*, *Formica cunicularia* (scouting species); *B*, *F. pratensis* (scroungers) in contact with *F. cunicularia*; *C*, *F. pratensis* without contact (Adapted from Reznikova 1982)

ants with food baits hidden in a "sectoral maze". Meat baits were offered in a certain area, the approach to which was divided into ten equal sectors (Fig. 3.3a). This apparatus was presented in plots visited by only one of the two species, and in plots visited by both. Improvements in the ease with which the bait was found over periods of exposure to the maze for 3 h for each experimental plot were recorded. The improvements of ants' actions were assessed by assigning error points, which were calculated as the sum of distances (measured in the number of sectors) between the sectors entered before the sector with the bait was reached and that sector. Thus, if an ant entered the sector opposite to the sector with the bait, and then visited all the sectors in the semicircuit of the apparatus on its way to the bait, it received $5+4+3+2+1=15$ error points. Individual marking showed that on each plot where the species were active a constant group of about ten *F. pratensis* and two to three *F. cunicularia* were working.

It turned out that in plots visited only by *F. cunicularia* within 1–15 min of the beginning of trials the bait was found by them and quickly dragged away. *F. cunicularia* acted individually. If they made a mistake, the ants ran around until they found the correct sector. This species did not need more than three or four visits to remember the right sector. On the contrary, in plots visited only by *F. pratensis*, these ants roamed about the sectors for an hour, or even two, and found the bait only by chance. Later on their searching became orderly, but the sum of errors remained high. If *F. pratensis* had the opportunity to observe *F. cunicularia*, then they did not touch the bait at all for the first 30–40 min, allowing the subdominants to drag it away, During this time *F. pratensis* foragers stood near the maze watching the actions of the scout species. Then they drove *F. cunicularia* away and collected the food pellets themselves (Fig. 3.3b).

In sum, in solely foraging species, unlike in mass recruiting species, each individual is smart. However, similarly with mass recruiting species, the solely foraging ones do not display any distribution of cognitive responsibilities within colonies.

3.3.3 Group-Retrieving Based on Cognitive Specialisation

The group-retrieving mode of foraging based on difference of cognitive responsibilities between scouts and foragers has been described in *Formica* s.str. (Reznikova and Ryabko 1994, 2011). Among many taxonomic groups, ants of the genus *Formica*, in particular, red wood ants (the *Formica rufa* group) are possibly the most promising and a surprisingly underestimated group for studying cognitive aspects of communication. In comparison with many sympatric species, red wood ants have hundreds of times more individuals per colony, as well as spacious feeding territories (Fig. 3.4). Every day they face more complex vital problems than many other species: for example, in order to obtain honey-dew, the basic food for adults, red wood ants have to find and possibly memorise locations of thousands of aphid colonies within such a huge three-dimensional space as a tree is for an ant; besides, each colony has to find, kill and transport thousands of invertebrates in order to feed larvae. Dobrzanska (1958) demonstrated that in red wood ants, groups of individuals return repeatedly to approximately the same parts of the colony's feeding territory and work together there. Studying site allegiance in red wood ants, Rosengren and Fortelius (1987) characterised red wood ants as "replete ants" storing not lipids in their fat-bodies but habitat information in their brains.

Although the *Fomica rufa* group is one of those at the centre of myrmecological studies, for a long time amazingly little has been known about their colony organisation and about the ways information is transmitted within colonies.

Perhaps the best model situation in which to observe task distribution and to evaluate potential properties of communication in group-retrieving ants is to force them to solve a complex search problem. In this situation the processes of information transmission and task allocation between members of working "teams" would become observable. These experiments are described in Chap. 4.

In general, social hymenopterans display high cognitive abilities at the individual level, which are comparable with those of monkeys and dogs. In ants, the distribution of cognitive responsibilities among individuals depends on which recruitment strategies they use, as well as on the level of social organisation within their family. Highly social species use distant homing, which, because of huge feeding territories and the group-retrieving mode of recruitment used by these species, is possibly the optimal way to transfer information about remote events. In Chap. 4 we consider how scouting ants pass "abstract" information to foragers.

Fig. 3.4 The ant-hill of red wood ants *Formica aquilonia* (Photo by Olga Sinkova)

References

Avarguès-Weber A, Giurfa M (2013) Conceptual learning by miniature brains. Proc R Soc Lond B Biol Sci 280(1772):20131907

Cammaerts Tricot MC (2012) Navigation system of the ant Myrmica rubra (Hymenoptera: Formicidae). Myrmecol News 16:111–121

Cammaerts MC, Cammaerts R (2015) The acquisition of cognitive abilities by ants: a study on three *Myrmica* species (Hymenoptera, Formicidae). Adv Stud Biol 7(7):335–348

Cerdá X, Angulo E, Boulay R, Lenoir A (2009) Individual and collective foraging decisions: a field study of worker recruitment in the gypsy ant Aphaenogaster senilis. Behav Ecol Sociobiol 63(4):551–562

Cherix D (1980) Preliminary note on the structure, phenology and diet of a supercolony of Formica lugubris Zett. Insect Soc 27(3):226–236

Collett TS, Fry SN, Wehner R (1993) Sequence learning by honeybees. J Comp Physiol A 172(6):693–706

Czaczkes TJ, Grüter C, Ellis L, Wood E, Ratnieks FL (2013) Ant foraging on complex trails: route learning and the role of trail pheromones in *Lasius niger*. J Exp Biol 216(2):188–197

Detrain C, Deneubourg JL, Pasteels JM (1999) Information processing in social insects. Springer Science & Business Media. Birkhauser Verlag, Basel
Dobrzanska J (1958) Partition of foraging grounds and modes of conveying information among ants. Acta Biol 18:55–67
Dornhaus A, Holley JA, Pook VG, Worswick G, Franks NR (2008) Why do not all workers work? Colony size and workload during emigrations in the ant *Temnothorax albipennis*. Behav Ecol Sociobiol 63(1):43–51
Dyer AG, Neumeyer C, Chittka L (2005) Honeybee (Apis mellifera) vision can discriminate between and recognise images of human faces. J Exp Biol 208(24):4709–4714
Evans S (1932) An experiment in maze learning with ants. J Comp Psychol 14(2):183
Fortelius W, Rosengren R, Cherix D, Chautems D (1993) Queen recruitment in a highly polygynous supercolony of *Formica lugubris* (Hymenoptera, Formicidae). Oikos:193–200
Giurfa M, Dafni A, Neal PR (1999) Floral symmetry and its role in plant pollinator systems. Int J Plant Sci 160(S6):S41–S50
Hahn M, Maschwitz U (1985) Foraging strategies and recruitment behaviour in the European harvester ant *Messor rufitarsis* (F.). Oecologia 68(1):45–51
Heinze J (2008) The demise of the standard ant (Hymenoptera: Formicidae). Myrmecol News 11:9–20
Higashi S, Yamauchi K (1979) Influence of a supercolonial ant Formica (Formica) yessensis Forel on the distribution of other ants in Ishikari Coast [Japan]. Jpn J Ecol 29:257–264
Hölldobler B, Traniello JF (1980) The pygidial gland and chemical recruitment communication in *Pachycondyla (= Termitopone) laevigata*. J Chem Ecol 6(5):883–893
Hölldobler B, Wilson EO (1990) The ants. Harvard University Press, Cambridge, MA
Jackson DE, Ratnieks FL (2006) Communication in ants. Curr Biol 16(15):R570–R574
Karas AY, Udalova GP (1999) Behavioral strategies in ants in changing conditions of food motivation. Neurosci Behav Physiol 29(3):263–269
Kartsev VM (2014) Situational choices among alternative visual stimuli in honeybees and paper wasps when foraging. In: Malloy C (ed) Honeybees: foraging behavior. Reproductive biology and diseases. Nova Science Publisher Hauppauge, NY, p 93–118
Leonhardt SD, Menzel F, Nehring V, Schmitt T (2016) Ecology and evolution of communication in social insects. Cell 164(6):1277–1287
Mazokhin-Porshnyakov GA (1969) Insect vision. Plenum Press, New York
Mazokhin-Porshnyakov GA, Kartsev VM (2000) Learning in bees and wasps in complicated experimental tasks. In: Prerational intelligence: adaptive behavior and intelligent systems without symbols and logic, volume 1, volume 2 prerational intelligence: interdisciplinary perspectives on the behavior of natural and artificial systems, vol 3. Springer, Netherlands, p 908–926
Mercier J-L, Lenoir A (1999) Individual flexibility and choice of foraging strategy in *Polyrhachis laboriosa* F. Smith (Hymenoptera, Formicidae). Insectes Sociaux 46(3):267–272
Miller N, Garnier S, Hartnett AT, Couzin ID (2013) Both information and social cohesion determine collective decisions in animal groups. Proceedings of the National Academy of Sciences 110(13):5263–5268
Novgorodova TA (2003) Intraspecific diversity of behavioral models in ants Formica cunicularia glauca under conditions of trophobiosis. Adv Curr Biol 123:229–233. (in Russian with English summary)
Reznikova ZI (1975) Non-antagonistic relationships of ants occupying similar ecological niches. Zool Zhurnal 54(7):1020–1031 (in Russian with English summary)
Reznikova Zh (1979a) Spatial orientation and logical skills in ants. In: Plehanov GF (ed) Ethology of insects and ticks. Tomsk State University, Tomsk, p 18–24. (in Russian)
Reznikova Zh (1979b) Different forms of territoriality in ants *Formica pratensis* Retz Zool J;58:1490–1499. (in Russian with English summary)
Reznikova Z (1980) Interspecies hierarchy in ants. Zool J 59:1168–1176 (in Russian with English summary)
Reznikova Z (1983) Interspecies relations in ants. Nauka, Novosibirsk, 205 pp. (in Russian)
Reznikova Z (1999) Ethological mechanisms of population density control in coadaptive complexes of ants. Russ J Ecol 30(3):187–192

Reznikova Z (2007a) Animal intelligence: from individual to social cognition. Cambridge University Press, Cambridge
Reznikova Z (2007b) Dialog with black box: using Information Theory to study animal language behaviour. Acta Ethologica 10(1):1–12
Reznikova Z (2008) Experimental paradigms for studying cognition and communication in ants (Hymenoptera: Formicidae). Myrmecol News 11:201–214
Reznikova Z (2011) Division of labour and communication at the individual level in highly social *Formica* ants (Hymenoptera, Formicidae). Russ Entomol J 20(3):315–319
Reznikova Zh (2012a) Intelligent communication in animals. In: Encyclopedia of the sciences of learning. Springer Verlag, Heidelberg, Germany, p 1597–1600
Reznikova Zh (2012b) Altruistic behavior and cognitive specialization in animal communities. In Encyclopedia of the sciences of learning. Springer Verlag, Heidelberg, Germany, p 205–208
Reznikova J(Zh.) (1982) Interspecific communication among ants. Behaviour 80;84–95
Reznikova Z, Ryabko B (1994) Experimental study of the ants communication system with the application of the Information Theory approach. Memorabilia Zoologica 48:219–236
Reznikova Z, Ryabko B (2003) In the shadow of the binary tree: of ants and bits. In: Anderson C, Balch T (eds) Proceedings of the 2nd Internat. Workshop of the mathematics and algorithms of social insects. Georgian Institute of Technology, Atlanta, pp 139–145
Reznikova Z, Ryabko B (2011) Numerical competence in animals, with an insight from ants. Behaviour 148(4):405
Rosengren R, Fortelius W (1986) Ortstreue in foraging ants of the *Formica rufa* group—Hierarchy of orienting cues and long-term memory. Insect Soc 33(3):306–337
Rosengren R, Fortelius W (1987) Trail communication and directional recruitment to food in red wood ants (*Formica*). In: Annales Zoologici Fennici. Finnish Academy of Sciences, Societas Scientiarum Fennica, Societas pro Fauna et Flora Fennica and Societas Biologica Fennica Vanamo. University of Helsinki, Finland, p 137–146
Schatz B, Lachaud JP, Beugnon G (1997) Graded recruitment and hunting strategies linked to prey weight and size in the ponerine ant *Ectatomma ruidum*. Behav Ecol Sociobiol 40(6):337–349
Schmid-Hempel P (1984) Individually different foraging methods in the desert ant *Cataglyphis bicolor* (Hymenoptera, Formicidae). Behav Ecol Sociobiol 14(4):263–271
Schneirla TC (1929) Learning and orientation in ants. Comparat Psychol Monogr 6:1–143
Smirnova A, Zorina Z, Obozova T, Wasserman E (2015) Crows spontaneously exhibit analogical reasoning. Curr Biol 25(2):256–260
Srinivasan MV (2010) Honey bees as a model for vision, perception, and cognition. Annu Rev Entomol 55:267–284
Stieb SM, Muenz TS, Wehner R, Rössler W (2010) Visual experience and age affect synaptic organization in the mushroom bodies of the desert ant *Cataglyphis fortis*. Dev Neurobiol 70(6):408–423
Thorpe WH (1950) A note on detour experiments with *Ammophyla pubescens* Curt. (Hymenoptera: Sphecidae). Behaviour 2:257–263
Vowles DM (1965) Maze learning and visual discrimination in the wood ant (*Formica rufa*). Br J Psychol 56(1):15–31
Wehner R (1972) Visual orientation performances of desert ants (*Cataglyphis bicolor*) toward astromenotactic directions and horizon landmarks. In book: Information Processing in the Visual Systems of Arthropods Springer Verlag Berlin Heidelberg, NY:295–302
Wehner R (2003) Desert ant navigation: how miniature brains solve complex tasks. J Comp Physiol A 189(8):579–588
Wehner R (2009) The architecture of the desert ant's navigational toolkit (Hymenoptera: Formicidae). Myrmecol News 12:85–96
Zhang SW, Lehrer M, Srinivasan MV (1999) Honeybee memory: navigation by associative grouping and recall of visual stimuli. Neurobiol Learn Mem 72(3):180–201
Zhang S, Bock F, Si A, Tautz J, Srinivasan MV (2005) Visual working memory in decision making by honey bees. Proc Natl Acad Sci U S A 102(14):5250–5255

Language Behaviour in Ants

4

4.1 Experimental Evidence of Information Transmission by Distant Homing in Ants

It had remained unclear for a long time whether ants can use distant homing for communication. As mentioned in Chap. 2, the so-called tactile (or antennal) "code" has been discussed since Wasmann (1899) first hypothesised the existence of distant homing in ants. Attempts to decipher ants' "tactile language", that is, to extract definite signals from their tactile communication, did not give the desired results. It was concluded that antennal movements have no structural unity of signals and replies (Lenoir 1982; Bonavita – Cougourdan and Morel 1984).

The early experimental results demonstrating ants' ability to transfer "abstract" information by means of distant homing were first obtained on *Camponotus herculeanus* (Reznikova 1979, 1983). In laboratory experiments ants had to transmit to each other information about food placed on one of 12 branches of one out of 10 artificial "trees" (Fig. 4.1). The scouting ants were allowed to visit the "trees" and to find food there, then they went to a separate part of the arena and made contacts with foragers in the nest. The scouts were then isolated for a while, and the "trees" were replaced in order to eliminate odour trails that could be laid by the scouts. Foragers successfully found the target after having had their contacts with scouts, whereas control ("uninformed") ants failed to do this. These experiments demonstrated that ants can transmit meaningful messages concerning the exact coordinates of a target. However, they could not be used to study characteristics of ant "language" beyond that.

The general experimental paradigm elaborated later in the "binary tree" study has been used then in all experiments on information transmission in ants (Reznikova and Ryabko 1986, 1994; Ryabko and Reznikova 1996, 2009). In these experiments ants were confronted with a rather complex life-or-death task: they could obtain food only in a "binary tree" maze and only once every 2–3 days. Ants therefore were hungry and extremely motivated to obtain some food. They had to search for the food placed on one of several "leaves" of the "binary tree" maze (Fig. 4.2). In

Zh. Reznikova, *Studying Animal Languages Without Translation: An Insight from Ants*, DOI 10.1007/978-3-319-44918-0_4

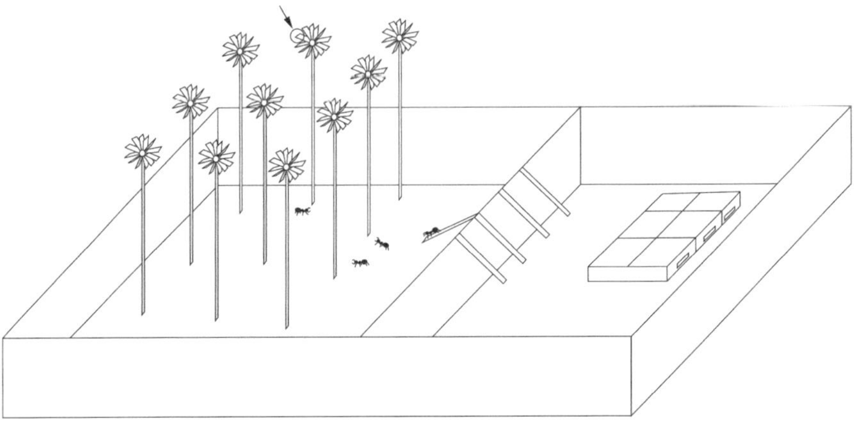

Fig. 4.1 The early experiment on information transmission in *Camponotus herculeanus*. Ants had to transmit to each other information about food placed on one of 12 branches of one out of 10 artificial "trees" (Adapted from Reznikova 1983)

various years three colonies of *Formica polyctena* and two of *F. sanguinea* were used, and we also analysed the results obtained by Novgorodova (2006) on *F. pratensis*. Ants lived in the 2 × 1.5 m laboratory arena, in a transparent nest that made it possible to observe their activity during experiments and was covered by a lid to maintain darkness the rest of the time. The arena was divided into two sections: a smaller one containing the nest, and a bigger one with the experimental system. The two sections were connected by a plastic bridge that was removed from time to time to modify the set-up or isolate the ants. To prevent access to the food in the maze by a straight path, the set-up was placed in a bath of water, and the ants reached the initial point of the binary tree by going over a second small bridge (Fig. 4.2). The laboratory colonies consisted of about 2000 individuals each. All actively foraging ants were individually marked with colour paint (Fig. 4.3).

In all series of experiments with the binary tree, the ants initially were able to feed on a one-fork set-up. In these cases, the foragers left their nest as a result of mass mobilisation. It was impossible to observe any structural units within the crowding ants. The ants' behaviour changed sharply when the trough with syrup was placed on one of the four leaves of the two-fork (two-level) binary tree, making their task more complicated. Then, in different series of experiments, binary tree mazes with different number of forks were used, from two to six (a maze on Fig. 4.2 has four forks). The leaf on which to place the filled trough was chosen randomly by tossing a coin for each fork in the path.

The laboratory colonies were found to include "teams" (see details in Sect. 4.2) of constant membership which consisted of one scout and four to eight recruits (foragers): the scout mobilised only members of its team to the food. The composition of the teams was revealed during special run-up experiments consisting of familiarisation trials lasting as long as 2 or 3 weeks. Ants freely visited the set-up and took syrup there. During this period a minority of scouts with poor records of

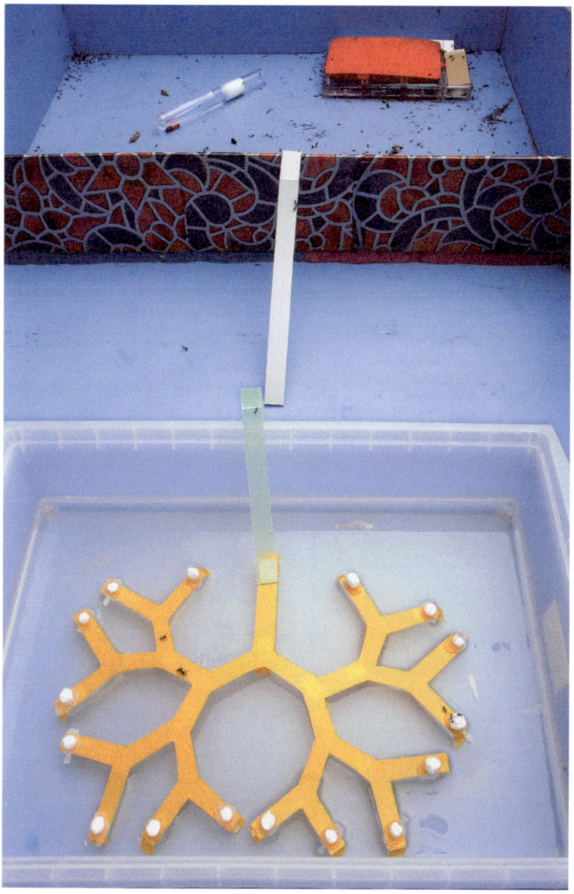

Fig. 4.2 A laboratory arena deviled into two parts, containing an artificial ant nest and a binary tree maze placed in a bath with water. This binary tree has four forks (Photo by Nail Bikbaev)

competence were identified and discarded. "Qualified" scouts and foragers belonging to their teams received special "team marks" by colour paint. Then the main series of experiments started. In total, 335 scouts along with their teams were used in all experiments with the binary tree. In each trial one of the scouts that were actively moving on the experimental arena at that moment was placed on a leaf of the binary tree that contained a trough with the food, and then it returned to the nest by itself. All contacts between the scout and its team were observed each time.

The scout had to make up to four trips before it was able to mobilise its group of foragers. Usually members of the team had already left the nest after the scout's first trip and were waiting on the arena for its return. Returning to the group, the scout contacted one to four foragers in turn, sometimes two of them simultaneously. It is worth to note again that the scout did not contact any other ants except members of its own group. Contacts were followed by numerous antennal movements (Fig. 4.4)

Fig. 4.3 Ants were individually marked with colour paint (Photo by Nail Bikbaev)

Fig. 4.4 A scouting ant contacting with members of its team (Photo by Nail Bikbaev)

(video records of some of the experiments are available at http://reznikova.net/research/ant-language/).

The experiments were designed to investigate the characteristics of distant homing, so after the scout had contacted its team, it was isolated for a while, and the foragers had to search for the food by themselves. All experiments were so devised as to eliminate all possible cues that could help the ants to find the food, except a contact with the scout. To avoid the use of an odour track, the experimental set-up was replaced by an identical one when the scout was in the nest or on the arena contacting its group (Fig. 4.5). All troughs in the fresh maze contained only water to avoid the possible influence of the smell of syrup. If the group reached the correct leaf of the binary tree, they were immediately presented with the food (Fig. 4.6 a, b).

Fig. 4.5 Dr. Natalia Atsarkina (our guest from Moscow State University) is marking ants by paint. One can see several extra mazes near the arena (Photo by Zhanna Reznikova)

It is important to note that the composition of the working teams remained constant in each colony from several days to several weeks, that is, during periods when a given scout was actively working. Moreover, not all of the scouts managed to memorise the way to the correct leaf of the maze even after they had passed their "final exams" during the run-up trials. The number of scouts that succeeded in memorising the way decreased with the increase in the complexity of the task. In the case of two forks all active scouts and their groups (up to 15 per colony) were successful, whereas in the case of six forks, only one or two coped with the task.

Each group was named after its scout (Roman numerals were used). The tasks were different on different days and for different groups. As an example, here is the seasonal history of the foraging groups of *F. sanguinea* used in the experiments (taken from: Reznikova and Ryabko 1994): group I was presented with the sequence of turns "LL", "LR", "LL" and "RR", "RL" and "LL", "LLL" and "LLR", "LRLL", "RRR", "LLLLL" on 14, 16, 18, 20, 22, 25 June and 2 and 5 July, respectively. Group VIII with "RLLLR", "LLL" and "RRR", "LRLL", "RRLRL" and "RRRRR", "LLLLL", "RRRRR" on 22, 25, 29 June and 2, 5 and 9 July, respectively. Note that

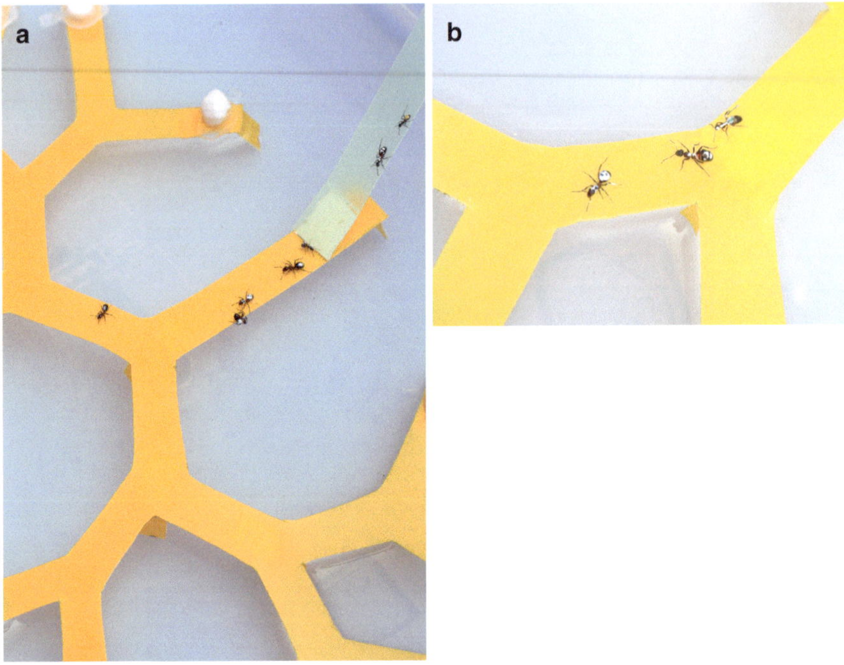

Fig. 4.6 A group of foragers descending the bridge (**a**) and traversing the maze (**b**) (Photo by Nail Bikbaev)

not all groups worked throughout the whole season, and sometimes new groups were observed to be active. To find and watch such groups several special tests with a simplified maze were carried out.

The evidence of information transmission from scouts to foragers comes from two sources: first, from the statistical analysis of the number of times the group found the target correctly, and second, from a special series of control experiments with "uninformed" ("naïve") and "informed" foragers.

The statistical analysis of the number of times of finding the target correctly was carried out by comparing the hypothesis H_0 (ants find the leaf containing the food by chance) with the hypothesis H_1 (they find the target thanks to the information obtained), proceeding from the fact that the probability of finding the correct way by chance when the number of forks is i is $(1/2)^i$. We analysed series of experiments separately for 2, 3, 4, 5 and 6 forks. For example, the probability of finding the correct way towards the trough in the maze with three forks by chance is 2^3. One of our series of trials with *F. sanguinea* on the binary tree with three forks included 20 trials. Out of these, in three cases the group of foragers failed to find the food; in 5 cases 1–3 ants were left behind the group and in 12 cases all foragers correctly reached the leaves where their scout had found syrup. A correct search was considered a "success", while the cases when the group failed to reached the target or not all of the members reached it was called a "failure." To perform the statistical

Table 4.1 Comparison of duration of searching the trough by "uninformed" (U) *F. pratensis* ants and individuals that previously contacted with the successful scout ("Informed", I)

Sequence of the turns	Ants (U/I)	Mean	Amounts of sampling	P
RRRR	U	345.7	9	<0.01
	I	36.3	9	
LLLL	U	508.0	9	<0.01
	I	37.3	9	
LRRL	U	118.7	7	<0.01
	I	16.6	7	
RLLR	U	565.9	7	<0.01
	I	16.3	7	

analysis, we have assumed that in this series of experiments we have the outcomes of 20 independent Bernoulli trials (with possible outcomes being "success" or "failure"), from which 12 "successes" and 8 "failures." The statistical analysis showed that in this case H_0 was rejected in favor of H_1, as well as in other series (338 trials in sum), thus unambiguously demonstrating information transmission from scouts to foragers.

The control experiments were organised so as to compare searching results of the ants that had and did not have a previous possibility to contact the scout (the "informed" and "naïve" ants, respectively). The "naïve" and "informed" ants were tested one by one. Each ant was allowed to search for the food for 30 min. For example, in one of the control experiments 26 uninformed *F. sanguinea* failed to find the food in the mazes with 4–6 forks and only one ant found the target by chance on the binary tree with 4 forks. In *F. polyctena* there were 2 successes versus 26 failures. In experiments with *F. pratensis* (Novgorodova 2006) the time spent on searching the trough by "informed" and "uninformed" specimens were compared (see Table 4.1). For every trial, Wilcoxon's non-parametric test was used to test the hypothesis H_0 (data from both samples follow the same distribution) against H_1 (they follow different distributions) at significance level 0.01. We concluded that the duration of search was essentially smaller in those ants that had previously contacted the scout. In sum, the difference between frequencies of findings of naive and informed ants, as well as the time spent for searching the target by them, were significantly different. The obtained data confirm information transmission by means of distant homing in three *Formica* species and exclude any orientation mechanism, except the use of information transmitted by the scouts.

It is worth noting that in the course of the binary tree study no evidence of distant homing was revealed in *F. cunicularia* or *Myrmica rubra* (Reznikova and Ryabko 1994, 2003). In the former species singly foraging species the foragers learned the way towards the maze, while making up to 30 trips per day, but they did not try to recruit other members of their colony. Not more than five ants were active in the maze per day. *M. rubra* workers utilise multiple recruitment system (Dlussky et al. 1978). In our experiment they used olfactory cues, but when we changed the maze, they had to do without odour trails. In these cases they resorted to only solitary foraging, just as *F. cunicularia*.

As noted before, some species with small colony size and a correspondingly small feeding territory can use tandem running for pair recruitment and odour trails for mass mobilisation to any point within their range. In those species that have large colonies and spacious feeding territories, such as red wood ants, often an ant hill is too far for a scout to attract inactive foragers or to lay a scent trail from. Distant homing is perhaps the most effective means of communication in this case. Foraging within a tree crown in F. polyctena is a good example here but it is also possible for ants to use distant homing in the steppe plains, which were demonstrated on *F. sanguinea* (Reznikova and Ryabko 1994, 2003) and *F. pratensis* (Novgorodova 2006).

4.2 Teams in Ants

In group-living animals division of labour is sometimes based on coordinated activities of group members. In relatively rare cases individuals form groups in which the members stay together for extended periods to accomplish a certain task. There are several examples of hunting teams in vertebrates (for a review Reznikova 2007a). Usually individuals coordinate efforts so that one or more individuals chase the prey, or flush it from hiding, while others head off its escape. The most organised groups in animal societies are based on a discrete division of labour that may be called "professional" specialisation. Many examples come from cooperative hunting (Packer and Ruttan 1988). For example, the bottlenose dolphins form groups of three to six. One dolphin acts as the "driver" and herds the fish in circles towards the "barrier dolphins" who are tightly grouped together to form a barrier. The driver performs fluke-slaps to cause the fish to leap into the air. As the fish begin to leap, the driver moves next to the barrier dolphins, who all catch the fish in the air with their mouths open (Gazda et al. 2005).

The discovery of teams in ants was connected with the discovery of the existence of complex communication system in group–retrieving ant species. As described in Sect. 4.1, such a communication system is based on scout–foragers informative contacts in which each scout transfers messages to a small constant group of foragers and does not pass the information to other groups. The ants thus work as coordinated groups which may be called teams.

We have not yet identified reliable behavioural signs in ants indicating personal recognition like in social wasps (Tibbetts 2002) and crayfish (Van der Velden et al. 2008); neither have we been able to train ants to distinguish between pictures of different individuals as in experiments with sheep (Kendrick et al. 2001), apes and monkeys (Parr et al. 2000), and domestic cattle (Coulon et al. 2009). Nevertheless, we can be confident on the ability of team members to recognise their "own" scout, and the ability of a scout to distinguish between its group and other nest mates. So we can consider ants team "personalised".

A field model of the situation in which ants act collectively to perform a specific task could be the organisation of honey dew collection by foragers of red wood ants in the tree crown. It is well known that ants look after symbiotic aphids, protect them from adverse conditions, and in return, ants "milk" the aphids, whose sweet excretions are one of the main sources of carbohydrate for adult ants (Mordvilko 1936; Bradley and Hinks 1968). In an ant family, there is a group of ants dealing with aphids (aphid tenders), which has a constant composition. Reznikova and Novgorodova (1998) described a system of intricate division of labour (professional specialisation) in aphid tenders. It turned out that in the species that exhibit the recruitment system based on distant homing (*F. polyctena* and *F. sanguinea*), each aphid colony is attended by a constant group (a team) of individuals which perform different tasks: "shepherds" only look after aphids and milk them, "guards" only guard the aphid colony and protect it from external threats, "transporting" ants transfer the food to the nest, and "scouts" search for new colonies (see Fig. 4.7). A given aphid tending "team" includes at least two to four shepherds, the same number of guards, and one scout. The professional specialisation increases the efficiency of trophobiosis. When ants were experimentally forced to change their roles, much food was lost. The ants belonging to the same aphid tending group distinguish at least two to three shepherds from two to three guards within this group (Reznikova 2007b, 2011; Novgorodova 2015). Such deep professional specialisation has only been found in the same species that exhibited the complex communication system in our studies on information transfer in ants (Reznikova and Ryabko 1994, 2001, 2003, 2011). For scouting ants the task in such a situation is to find an aphid colony, inform other ants about the new food source, and organise honey dew collection and transportation. It is difficult to observe how a scout attracts its team to a new aphid colony in the wild, and we succeeded in doing this only twice. In both cases the scouts found a female aphid that had established a new colony and then by means of distant homing recruited members of their teams who were waiting at another branch of the same tree (Reznikova 2007a).

The process of team formation in *Formica* has been monitored during an ontogenetic study (Reznikova and Novgorodova 1998; Reznikova 2008). We observed the ontogenetic trajectories of 80 newly hatched *F. sanguinea* in one of the laboratory colonies and watched the shaping of teams. There were 16 working teams in that colony which mastered binary tree mazes. From 80 individually marked naïve ants, 17 entered seven different working teams, with one to four individuals joining each. Three became scouts, two of them starting as foragers on joining two different teams and one starting as a scout at once. The age at which the ants were capable of taking part in the working groups as foragers ranged from 18 to 30 days, and those three ants which became scouts were aged 28–30 days.

Team fidelity was examined in two colonies of *F. sanguinea* and *F. polyctena*. In a separate experiment we removed foragers of nine teams (we moved them to another laboratory nest) thus leaving nine scouts alone in a common experimental colony (Reznikova and Novgorodova 1998). Of those nine scouts three managed to attract new foragers, four scouts were working alone, and two ceased to appear in the arenas. In another experiment scouts from five *F. polyctena* teams were removed,

Fig. 4.7 "Professional specialisation" within an ant team working on an aphid colony: two "guards" are actively protecting aphids from an enemy (a tufted tit) while "shepherds" are continuing with aphid milking, and a transporting ant is descending along a twig loaded with honeydew (Photo by Nail Bikbaev)

and team members were left alone in a common experimental colony. Foragers from these groups sometimes appeared in the arena. We placed 15 of these foragers one by one on the leaves of the binary tree maze containing a trough with food. After they returned to the nest they made only a few occasional contacts with other ants, and did not try to form teams.

All these results in sum show that in group-retrieving *Formica* species colonies contain scout-foragers teams of constant membership in which distant homing is based on direct exchange of information between a scout and its foragers. Each scout shares information on the discovered food only with its own group of foragers (4–8 individuals).

Formation of teams is apparently a complex process which is based on intricate long-term relations between scouts and foragers.

4.3 What Is It Like to Be a Scouting Ant?

The club of animals that use distant homing for transferring symbolic information includes highly social species of ants and bees. We know now that within families of these insects a small percentage of out-nest individuals act as scouts and search for new sources of food. Characteristic features of members of these "cognitive elites" remained enigmatic for a long time. It is known that scouting bees constitute a very specific group in a hive. Recently some of the molecular underpinnings of their behavior relative to foragers have been revealed (Liang et al. 2012). However, unlike ants, bees do not transfer information only to the members of their own team but rather to anyone that is interested. There is no team fidelity within foraging and scouting bees (for details, see Tautz 2008). We hypothesized that scouting ants, with their amazing team fidelity and ability to memorise and transfer exact information about coordinates of objects possess eclectic behavioural features. In order to reveal these characteristics, we elaborated a special battery of behavioural tests (Iakovlev 2010; Atsarkina et al. 2014).

First, we carried out a casting process in order to select scouts and members of their foraging teams with the use of the "binary tree" paradigm. Similarly with experiments described in Sect. 3.3, the ants were fed only once per two or three days and only in the "binary tree" maze. During the "foraging session" we let 50 ants to visit the working part of the arena, and then we removed the bridge which connected the working part of the arena with the living one in order to prevent other ants from visiting the maze.

We considered an ant scout if she:

1. Found the feeder first
2. Memorized its location without the possibility to use the odour track
3. Having collected the syrup returned to the nest immediately and directly
4. Entered into antennal contacts with other ants
5. Made repeated trips between the nest and the maze.

Out of the ants feeding in the maze we selected those that spent not more than 5 min on the trough and then hurried on towards the nest in a beeline. Manipulating with the bridge, we let them to move from the maze to the nest. After two or three trips some of those ants succeeded to mobilise members of their teams that were also let to visit the maze. Periodically we replaced the maze on the arena with a fresh one in order to prevent ants from laying the odour trail. After 5–6 "foraging sessions" we thus could select scouting ants and members of their foraging teams (let us call them "foragers" here).

We considered ants "foragers" if they:

1. Visited the feeder after their contacts with a scout without the possibility to use an odour track
2. Did not repeat their trips and were not prone to contact others

In sum, we selected 23 scouts and 54 foragers during two years of working with two laboratory colonies. It is worth to note that only about 3 % of out-nest workers act as scouts (Reznikova 2007a, 2011), so the number of scouts selected in our experiments was close to the limit for laboratory colonies. As a control group we used randomly selected active out-nest workers not mobilised by scouts. We also selected members of other task groups. Aphid milkers were identified as those members of the basic laboratory colony that attended to the aphids on an aspen branch placed on the arena. Guards were identified as those members of the laboratory colony that were on duty on the lid of the laboratory nest and reacted aggressively when approached with a needle. Hunters were identified as those ants that caught prey and transported it to the nest. All ants were individually marked with colour paint.

To reveal distinctive features of scouts in comparison with members of other task groups we firstly designed a battery of behavioural tests:

1. To specify exploratory activity we applied the experimental paradigm "a piece of artificial world" firstly presented by Reznikova (1983, 2007a) in the field study of exploratory activity in different ant species. The main idea was to estimate and compare the time duration which animals spend on models that simulate pieces of nature. In nature ants usually search for food in shelters, under stones, within grass columns and so on. Each ant knows all the ins and outs of its territory, and the introduction of new things triggers its exploratory behaviour. In laboratory tests we presented ants one by one with models which imitate natural situations, such as shelter ("litter setup"), tree trunk, and grass stems imitated by "brushes" made of plaster bars (Fig. 4.8 a, b).
2. We measured "aggressiveness" in dyadic interactions of an ant and a ground beetle. This method was first presented by Reznikova and Dorosheva (2004, 2013). They demonstrated ground beetles of the genus *Pterostichus* as "dear

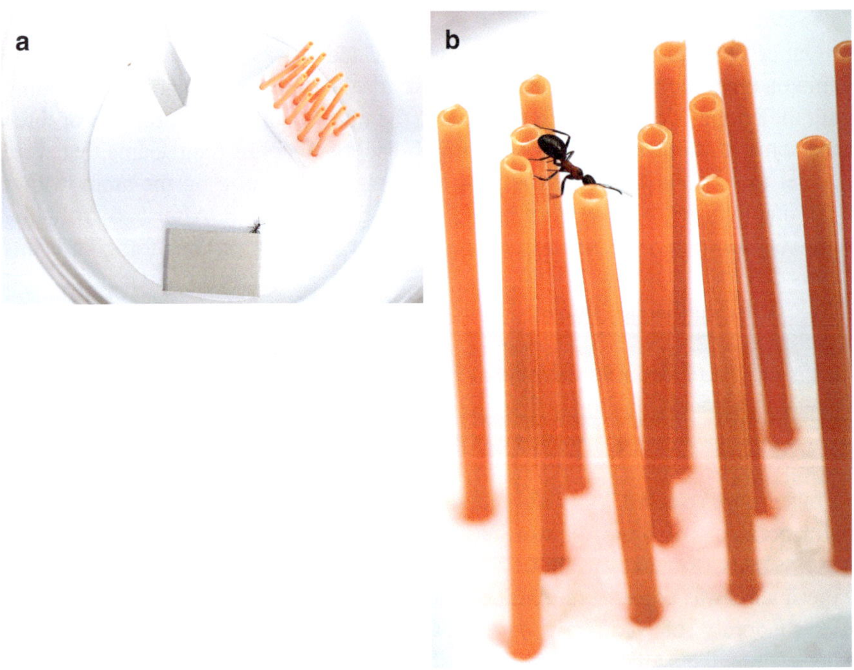

Fig. 4.8 A piece of "artificial world" designed for studying ants' exploratory activity. (**a**) A shelter ("litter setup"), tree trunk, and grass stems imitated by "brushes" made of plaster bars (an ant is climbing out of a shelter); (**b**) an ant is exploring "grass" (Photo by Nail Bikbaev)

enemies" for red wood ants, which always trigger ants' aggressive behaviour. We placed an ant and a beetle into a small arena (15×15×2 cm) and recorded all behavioural reactions during 10 min (Fig. 4.9a–d).

3. To estimate and compare cognitive abilities of scouts and foragers we used the binary tree again; however, the task was not to find food but to memorise the pass to the nest.

Comparison of exploratory behaviour in members of different task groups has shown that scouts and foragers are more mobile than members of the control group; scouts prefer grass setup, and they are not prone to hide under the shelter. By the type of exploratory activity scouts are most close to aphid milkers (Fig. 4.10). Scouts as a group behave more uniformly than others: all scouts investigate all elements of the artificial world, whereas only half of both foragers and members of the control group display any interest at all: they hide under the shelter or simply stay put.

It is interesting to note that those foragers and control ants who displayed the interest to the elements of an artificial world investigated these elements longer than an "average" scout. The most characteristic feature of the scouts is that they switch between activities more frequently than foragers and members of the control group (Fig. 4.11).

Fig. 4.9 Aggressive reactions of ants and ground beetles (*Pterostichus magus*) towards each others: (**a**) an ant is pursuing a beetle with open mandibles; (**b**) an ant is biting a beetle trying to turn it over and kill; (**c**) an ant is killing a beetle; (**d**) a beetle is killing an ant (Photo by Nail Bikbaev)

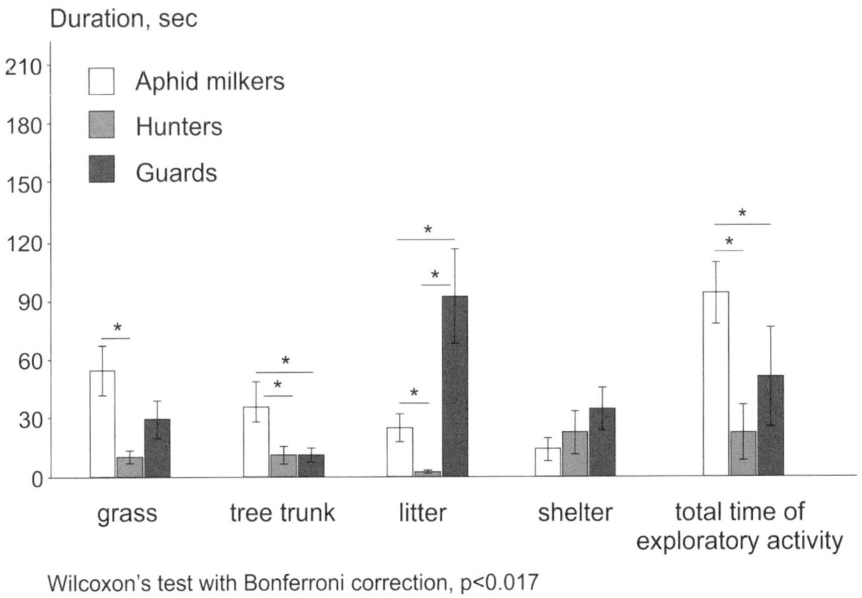

Fig. 4.10 By the type of exploratory activity scouts are close to aphid milkers

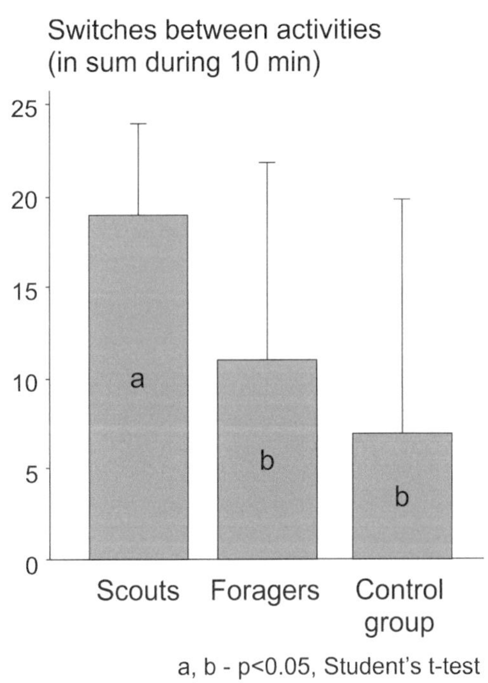

Fig. 4.11 The most characteristic feature of the scouts is that they switch between activities more frequently than foragers and members of the control group

All ants to a lesser or greater extent displayed aggressive reactions towards their "dear enemies", that is, the ground beetles (see details in: Reznikova 2007a, b). There is a variety of displays of attacks of different levels of aggressiveness, such as lunge, aggressive inspection, pursuit, bite and "mortal grip," which is rather suicidal and can end with the ant being killed (Fig. 4.9d). By the level of aggressiveness scouts and foragers are somewhere between aggressive guards and hunters on the one extreme and peaceful aphid milkers on the other extreme (Fig. 4.12). Specifics of aggressive behaviour in scouts is that they inspect the enemy in different ways and strongly avoid clinch fighting. In comparison with members of other task groups, scouts share more features of aggressive behaviour with aphid milkers than with guards and hunters.

Basing on the whole body of data that we obtained in the "binary tree" study as well as in the "counting maze" study (Reznikova and Ryabko 1994, 2011), we can say that scouts learn faster and remember better than foragers.

In sum, what it is like to be a scout we do not know yet. However, our new battery of tests enabled us to reveal character features of these members of the intellectual elite of the ant-hill. They are the following.

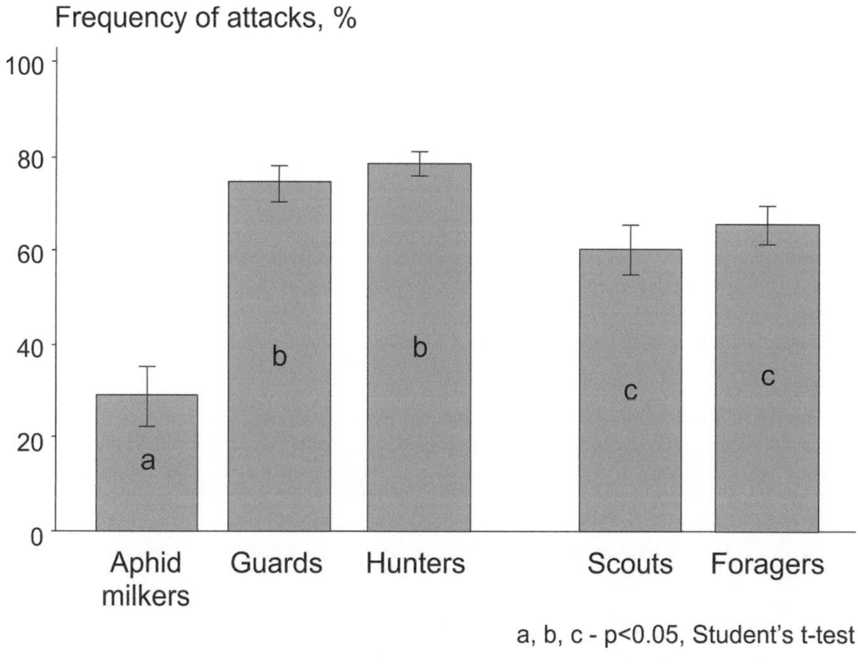

Fig. 4.12 By the level of aggressiveness scouts and foragers are somewhere between aggressive guards and hunters on the one extreme and peaceful aphid milkers on the other extreme

1. Scouts are explorative. Both scouts and members of their foraging groups are more explorative than "average" out-nest workers, and scouts are more explorative than foragers. Scouts share features of exploratory activity with aphid milkers: they display a higher interest to artificial grass and tree trunks. The most characteristic features of scouts are a high frequency of switching between activities and faithful interest to the variety of stimuli (one can say that scouts suffer from ADHD).
2. Scouts are brave. Aggressiveness in scouts and foragers is higher than in aphid milkers but lower than in guards. In contrast to guards, both scouts and foragers strongly avoid aggressive actions which may be of danger to themselves, and they never attack the enemy directly.
3. Scouts are smart. We can suggest that they form spatial memory faster and keep the information longer and better than foragers.

In conclusion, in red wood ants only scouts can do clever things such as remembering sequences of turns towards the target, noting regularities and even doing simple arithmetic, as we will see further. What is even more interesting, this is their lifelong role.

So when you accidentally crush an ant, it may mean nothing for the interaction network of which it was part, but it is possible that you have deprived a community of a member of an elite club of rare cognitive specialists.

At last, we now have members of Formica *rufa* group as the most reliable agents for studying animal language without translation, on the basis of the information theory approach. These experiments are described in Chaps. 5 and 6.

References

Atsarkina N, Iakovlev I, Reznikova Z (2014) Individual behavioural features of scouts and recruits in red wood ants (Hymenoptera: Formicidae). Euroasian Entomol J 13(3):209–218. in Russian with English summary

Bonavita – Cougourdan A, Morel L (1984) Les activites antennaires au cours des contacts trophallactique chez la Fourmi *Camponotus vagus* Scop. Ont-elles valeur de signal? Insect Soc 31:113–131

Bradley GA, Hinks JD (1968) Ants, aphids, and jack pine in Manitoba. Can Entomol 100:40–50

Coulon M, Deputte BL, Heyman Y, Baudoin C (2009) Individual recognition in domestic cattle (Bos taurus): evidence from 2D-images of heads from different breeds. Plos One 4(2):e4441

Dlussky GM, Volcit OV, Sulhanov AV (1978) Organization of group foraging in ants of the genus *Myrmica*. Zool J 57:65–77 (In Russian with English summary)

Gazda SK, Connor RC, Edgar RK, Cox F (2005) A division of labour with role specialization in group–hunting bottlenose dolphins (*Tursiops truncatus*) off Cedar Key, Florida. Proc R Soc Lond B Biol Sci 272(1559):135–140

Iakovlev IK (2010) Ethological aspects of task specialization in red wood ant (Hymenoptera: Formicidae) colonies. Proc Russ Entomol Soc 81(2):180–187

Kendrick KM, da Costa AP, Leigh AE, Hinton MR, Peirce JW (2001) Sheep don't forget a face. Nature 414(6860):165–166

Lenoir A (1982) An informational analysis of antennal communication during trophallaxis in the ant *Myrmica rubra* L. Behav Processes 7(1):27–35

Liang ZS, Nguyen T, Mattila HR, Rodriguez-Zas SL, Seeley TD, Robinson GE (2012) Molecular determinants of scouting behavior in honey bees. Science 335(6073):1225–1228

Mordvilko AK (1936) Ants and aphids. Priroda 4:44–55. (In Russian)

Novgorodova TA (2006) Experimental investigation of information transmission in *Formica pratensis* (Hymenoptera, Formicidae) using "binary tree" maze. Entomol Rev 86(3):287–293

Novgorodova TA (2015) Role of social and individual experience in interaction of the meadow ant *Formica pratensis* (Hymenoptera: Formicidae) with ladybird imagines and hoverfly larvae. Insect Sci 22(3):440–450

Packer C, Ruttan L (1988) The evolution of cooperative hunting. Am Nat 132(2):159–198

Parr LA, Winslow JT, Hopkins WD, de Waal F (2000) Recognizing facial cues: individual discrimination by chimpanzees (*Pan troglodytes*) and rhesus monkeys (*Macaca mulatta*). J Comp Psychol 114(1):47

Reznikova Zh (1979) Spatial orientation and logical skills in ants. In: Plehanov GF (ed) Ethology of insects and ticks. Tomsk State University, Tomsk, p 18–24. (in Russian)

Reznikova Z (1983) Interspecies relations in ants. Nauka, Novosibirsk, 205 pp. (in Russian)

Reznikova Z (2007a) Animal intelligence: from individual to social cognition. Cambridge University Press, Cambridge

Reznikova Z (2007b) Dialog with black box: using Information Theory to study animal language behaviour. Acta Ethologica 10(1):1–12

Reznikova Z (2008) Experimental paradigms for studying cognition and communication in ants (Hymenoptera: Formicidae). Myrmecol News 11:201–214

Reznikova Z (2011) Division of labour and communication at the individual level in highly social *Formica* ants (Hymenoptera, Formicidae). Russ Entomol J 20(3):315–319

Reznikova Z, Dorosheva H (2004) Impacts of red wood ants *Formica polyctena* on the spatial distribution and behavioural patterns of ground beetles (Carabidae). Pedobiologia 48(1):15–21

Reznikova Z, Dorosheva E (2013) Catalog learning: carabid beetles learn to manipulate with innate coherent behavioral patterns. Evol Psychol 11(3):147470491301100304

Reznikova Z, Novgorodova TA (1998) Individual task allocation and information transmission in ants' working groups. Adv Curr Biol 118:345–356. (in Russian with English summary)

Reznikova Z, Ryabko B (1986) Investigations of ant language by methods of Information Theory. Probl Inf Transm 21:103–108

Reznikova Z, Ryabko B (1994) Experimental study of the ants communication system with the application of the Information Theory approach. Memorabilia Zoologica 48:219–236

Reznikova Z, Ryabko B (2001) A study of ants' numerical competence. Electron Trans Artif Intell B 5:111–126

Reznikova Z, Ryabko B (2003) In the shadow of the binary tree: of ants and bits. In: Anderson C, Balch T (eds) Proceedings of the 2nd Internat. Workshop of the mathematics and algorithms of social insects. Georgian Institute of Technology, Atlanta, pp 139–145

Reznikova Z, Ryabko B (2011) Numerical competence in animals, with an insight from ants. Behaviour 148(4):405

Ryabko B, Reznikova Z (1996) Using Shannon Entropy and Kolmogorov Complexity to study the communicative system and cognitive capacities in ants. Complexity 2:37–42

Ryabko B, Reznikova Z (2009) The use of ideas of information theory for studying "language" and intelligence in ants. Entropy 11(4):836–853

Tautz J (2008) The buzz about bees: biology of a superorganism. Springer Science & Business Media. Springer, Heidelberg

Tibbetts EA (2002) Visual signals of individual identity in the wasp *Polistes fuscatus*. Proc R Soc Lond B Biol Sci 269(1499):1423–1428

Van der Velden J, Zheng Y, Patullo BW, Macmillan DL (2008) Crayfish recognize the faces of fight opponents. PLoS One 3(2):e1695

Wasmann E (1899) Die psychischen Fähigkeiten der Ameisen. Zoologica 26:1–133

Information-Theoretic Methods for Studying Ant "Language"

5

It is a natural idea to use an information-theoretic approach to investigate animal language behaviour, because information theory provides general principles and methods for studying and developing effective and reliable communication systems. Shannon (1948) established the basis of information theory in the end of the 1940s. The fundamental role of this theory was appreciated immediately, not only in the technology of information transmission, but also in the study of natural communication systems including human languages (Shannon 1951; Yaglom and Yaglom 1983).

Most of the literature that uses information-theoretic methods to study animal communications represents the signals emitted by animals as "texts". Thus, a growing body of literature attempts to analyse the information content of vocal signals of various species of birds, as well as whales, dolphins, monkeys and some others (for reviews, see: Bradbury and Vehrencamp 1998; McCowan et al. 1999, 2002). As far as we know, before our ant studies there were no experimental paradigms for direct examination of such properties of natural communication as the rate of information transmission, productivity and flexibility. We have focused on the conceptually distinct method based on ideas of information theory and aimed at investigation of the very process of exact information transmission between individual agents. The results are described further in this Chapter.

5.1 Basic Information-Theoretic Notions Used to Study Animal Communications

A central object of study in Shannon's information theory is the length of a message that one would need in order to communicate (transmit or record) the outcome of an uncertain event. An important measure of information is known as the *entropy* (more specifically, Shannon entropy), that is, the expected average value of the information contained in each message. Shannon entropy is usually expressed by the average number of bits needed to store one symbol in a message.

Zh. Reznikova, *Studying Animal Languages Without Translation: An Insight from Ants*, DOI 10.1007/978-3-319-44918-0_5

One *bit* is the information content of an event when the probability of that event occurring is one half. Consider a "heads-or-tails" trial with a "fair" coin, that is, a trial where both possible outcomes have equal probabilities. To record the outcome of this trial, we can write 1 for "tails" and 0 for "heads", or vice-versa; in any case, the length of the message is 1. The uncertainty of this trial is one bit, and if somebody informs us about the outcome, he or she transfers one bit of information. In general, if a trial has n equiprobable outcomes, then by being informed about its result we receive $log_2 n$ bits of information.

Imagine that you know that that a friend lives in a house with two entrances, two storeys and two flats on each storey, and you do not know in which flat your friend lives. There are 8 flats, and suppose that the probability of finding your friend in anyone of them equals 1/8. If somebody informs you at which flat you should call, he or she transfers $log_2 n = 3$ *bits* of information. In this case it is easy to follow each bit: information about one of the two entrances, then about one of the two storeys, and last about one of the two flats. If you agree in advance how to communicate where your friend lives, this could be done with 3 binary symbols, for example 001 or 110. One could use longer messages, but, on average over the 8 possible outcomes, one could not do with less than 3 binary symbols. Thus, 3 is the minimal expected length of the message necessary to communicate the random outcome.

More generally, if there are n possible outcomes that do not have to be equiprobable but rather have probabilities p_i (where i is from 1 to n), the shortest expected length of a message necessary to communicate the outcome is given by $\sum_{i=1..n} p_i \log_2 1/p_i$. This is, by definition, its *entropy*. Shannon has demonstrated that this expected length is indeed achievable (up to a tiny constant) using a certain code.

Note that, in order to measure the length of a message one does not need to understand its meaning. There may be many different codes that give the same expected length of the message for our stochastic trial in question. However, we do not need to know the code in order to judge how efficient it is: we can compare the average length of the encoded messages with the minimum possible, and, as described above, to calculate this minimum possible expected length all we need to know is the probabilities of all possible outcomes.

In real life, of course, rarely do we know the probabilities of all the outcomes. What is the probability that someone lives at a specific address in a given city? What is the probability that a new food source for an ant family is located at this specific branch of a tree? We have no means to reason precisely about this. However, if we design an experiment where we place the food where we want, then it is up to us to specify whatever probabilities we want, and then select the food location accordingly. This is what is done in the experiments described in the following sections.

However, there is one more important obstacle that presents itself on this path. Namely, measuring the length of a message in natural communications is anything but easy. Imagine you hear someone talking over a phone in an unknown language. Imagine also that you know exactly what information the person is trying to communicate. What is the length of the sentence? Of course, without knowing the language we do not know. We can, however, measure the time it takes the person to pronounce it. Knowing the amount of information being transmitted, and the time it takes to transmit, means that we know the rate of information transmission (bits per second). As shown below, this measure alone, used in a series of carefully designed experiments, allows us to judge not only about the efficiency of animal communications, but also about such intellectual abilities of the subjects as their numerical competence (see Chap. 6).

While Shannon's probabilistic notion of information and entropy is already very useful for studying natural communication systems, its important limitation is that it does not allow one to reason about information content of an individual object, that is, of a single message that is not one of a variety of messages produced at random. To overcome this shortcoming, the notion of algorithmic information that became known as *Kolmogorov complexity* has been proposed (Kolmogorov 1965). Here as well, the amount of information in an object may be interpreted as the length of a description of the object. More precisely (but still informally), the Kolmogorov complexity of a message is the length of the shortest program (on a certain reference computer) that generates this message and then halts. It can thus be thought of as the length of the maximally compressed version of the message (for details see: Li and Vitányi 2009).

Intuitively, it seems to us that the string 000000000000000 is much simple than, say, 011010101110010O, even though they have the same length. The first one seems more "regular:" it can be expressed as "15 zeros." One can imagine this as a short program for a computer: "print 15 zeros". The second string appears "random:" probably the shortest computer program to produce it just has to record the whole string explicitly. It is easy to see that, at least for some lengths of strings, the first program is much shorter. The notion of Kolmogorov complexity formalizes this intuition: more regular objects are those produced by shorter programs. Random objects are those for which the shortest program is just a one that lists the whole object explicitly, and thus has roughly the same length as the object itself. The main limitations of this approach is that Kolmogorov complexity of an object is not algorithmically computable. While one can obtain some estimates of it using data compressors (Cilibrasi and Vitányi 2005; Ryabko et al. 2006, 2013), in this book we do not use more than the intuition behind this approach to say that some individual messages, which are asking our ants to transmit, are more complex than others.

5.2 Shannon Entropy and Information Transmission in Ants

The idea of our experiments with ants is that we know exactly the quantity of information to be transferred, and can measure the time that ants spend doing it. To present ants with situations of equiprobable choices, we used the "binary tree" maze with different numbers of forks. The "binary tree" maze is described in Chap. 4, in which the evidence of information transmission from scouts to foragers by means of distant homing is presented. In this section we investigate the key characteristics of ant "language" such as rates of information transmission and potential numbers of messages these insects can produce.

The binary tree maze presents a situation in which, in order to obtain food, ants have to transmit the information about the sequence of turns towards a target. The target is a trough filled with syrup, located on one of the leaves of a binary tree; all the other leaves of the tree contained empty troughs. The leaf on which to place the filled trough was chosen randomly by tossing a coin for each fork in the path. The simplest design is a tree with one fork and two leaves, that is, a Y-shaped maze (Fig. 5.1 shows the trees with one and with four forks). It represents one binary choice which corresponds to one bit of information. In this situation a scouting animal should transmit one bit of information to other individuals: to go to the right (R) or to the left (L). In other experiments the number of forks of the binary tree increased up to six. Since the leaf on which to put the trough with syrup was chosen uniformly at random, the number of bits necessary to choose the correct way is equal to the number of forks, that is, the number of turns to be taken (Fig. 5.2 shows a maze with 3 forks).

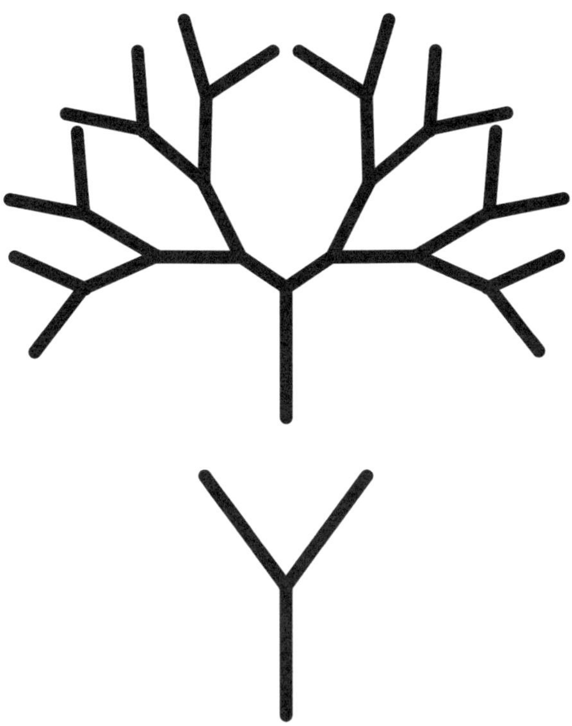

Fig. 5.1 The maze "binary tree" with one fork and four forks

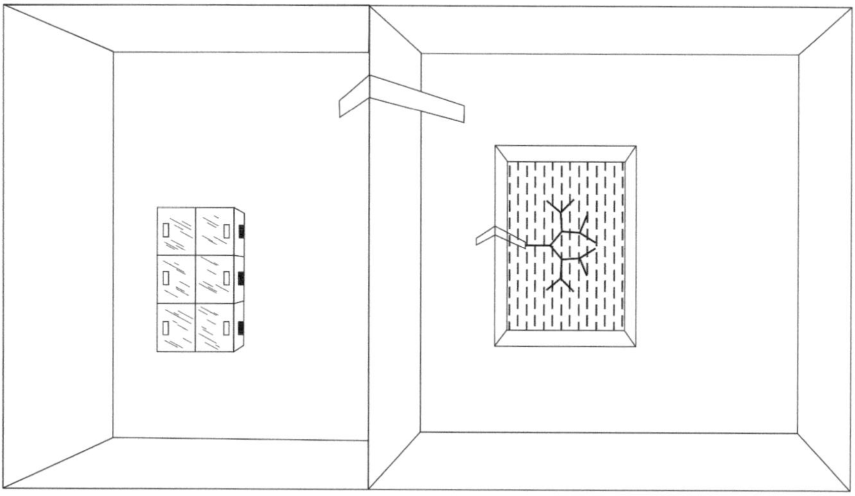

Fig. 5.2 The laboratory arena with the maze "binary tree" (a scheme)

As mentioned in Sect. 4.1, the use of an odour trail was excluded by replacing the whole set-up with a fresh one in the course of each trial. During the experiments each scout was placed on the trough containing food, and, after the scout had eaten, it returned to the nest on its own. Sometimes the scout contacted its team at once, and the group promptly began to move towards the set-up. In this case, after the scout contacted the foragers it was removed for a while, and the foragers had to search for the food by themselves. But more often the scout turned to go back to the trough alone. Sometimes it made errors and found the trough containing food only after visiting some empty troughs. Finally, after a number of visits it returned to the nest again and made a contact with its team that lasted longer than before, mobilising foragers to go to the maze. Sometimes the scout had to make up to four trips before it could mobilise the foragers. In all cases of mobilisation the duration of the contact between the scout and the foragers was measured (in seconds). The contact was considered to begin when the scout touched the first forager ant, and to end when the first two foragers left the nest for the maze. When the scout repeatedly returned to the trough alone, each of its contacts with foragers was measured. However, only the duration of the contact after which the foragers left the nest was taken into account. These contacts were hypothesised to be "informative" (see Fig. 4.4 in Chap. 4).

Evaluation of information transmission rate in ants is based on the fact that the quantity of information (in bits), necessary for choosing the correct way toward the maze, equals i, the depth of the tree (the number of turns to be taken), that is, $\log_2 n$ where n is the number of leaves of the binary tree. One can assume that the duration of the contacts between the scouts and foragers (t) is $ai+b$, where i is the number of turns (the depth of the tree), a is the rate of information transmission (bits per minute), and b is a constant added to account for the information ants may transmit that is not related directly to the task, for example, the simple signal "food". Besides, it is not ruled out that a scout ant transmits, in some way, the information on its route to the nest, using acoustic or some other means of communication. In this context, it is important to note that the route from the maze to the nest on the arena was in all experiments approximately the same, that is, the experimental setup was always placed on the same part of the working part of the arena. Thus, the amount of information potentially transmitted on the way to the nest would be roughly constant, and thus is accounted for in the constant b. Being highly motivated (see Sect. 4.1 for details), scouts hurried on to the nest in a beeline, and, therefore, the time before they made antennal contact with the foragers in the nest, which the scout could hypothetically use for message transmission, was approximately the same and did not depend on the number of turns to be taken in the maze.

The rate of information transmission (a in the formula above) was found to be about 1 min per bit in all the three ant species studied, which is at least ten times smaller than in humans.

One of the important properties of natural languges is "productivity," that is, the ability for generating a potentially unlimited number of messages on the basis of finitely many signals. To estimate the productivity of ants' "language," let us count the total number of different possible routes to the trough. In the simplest binary tree with one fork there are two leaves and therefore two different routes. In a tree with two forks there are 2^2 routes, with three forks 2^3 routes, and with six forks, 2^6 routes; hence, the total number of different routes is equal to $2+2^2+2^3+\ldots\ 2^6=126$. This is the minimal number of messages the ants must possess in order to pass the information about the food placed on any leaf of the binary tree with 6 forks.

5.3 Kolmogorov Complexity and Data Compression in the Ants' "Language"

Another series of experiments with the binary tree was inspired by the concept of Kolmogorov complexity and was designed to check whether ants possess such an important property of intelligent communications as the ability to grasp regularities and to use them for encoding and "compressing" information. This concept can be applied to words (or text) composed of the letters of any alphabet, for example, of an alphabet consisting of two letters: L and R. We interpret a word as a sequence of left (L) and right (R) turns in a binary tree maze.

Informally, the Kolmogorov complexity of a word is the lengh of its most concise description. For example, the word "LLLLLLLL" can be represented as "8 L", the word "LRLRLRLR" as "4LR", while the "random" word of shorter length "LRRLRL" probably cannot be expressed more concisely, and this is the most complex of the three.

We analysed the question of whether ants can use simple regularities of a sequence, such as those above, in order to compress it. As noted previously, Kolmogorov complexity is not computable. Therefore, we use it more as an inspiration than as a formal notion. What we can say about the strings in the example of the previous paragraph is that some of them appear to us to be more complex than others. In our experiments with the binary tree maze, different sequences of turns have different complexities, and we attempt to see whether the same sequences that appear more complex to us also appear more complex to ants.

In the course of a series of experiments, *Formica sanguinea* were presented with different sequences of turns. Different ant groups mastered each sequence in several trials (from 4 to 18), and we measured the time each scout spent transferring the information about each sequence of turns. These data are presented in Table 5.1. Evidently, most people perceive the sequence of turns given on the 5th (LLLLLL) and 6th (RRRRRR) lines of the table as simpler than those on the 7th (LRLRLR) and 8th (RLRLRL) lines, which, in turn, are simpler than the random ones of the same length given on the lines 13–15 (such as LRLLRL). As can be seen in Table 5.1, the same holds for ants: those sequences that look simpler to are apparently simpler to them, since the time the ants spent transmitting the information about the sequence of turns increased with its complexity. This means that the ants are able to grasp those regularities in these sequences that we see.

More formally, we applied a statistical test in order to examine whether the time for transmission of information by ants depends on its complexity (Reznikova and Ryabko 1994; Ryabko and Reznikova 2009). We considered two hypotheses: H_0

Table 5.1 Duration of transmitting information on the way to the trough by *F. sanguinea* scouts to foragers (no. 1–8 regular turn pattern; no. 9–15 random turn pattern)

No	Sequences	Mean duration (sec.)	SD	Numbers of experiments
1	LL	72	8	18
2	RRR	75	5	15
3	LLLL	84	6	9
4	RRRRR	78	8	10
5	LLLLLL	90	9	8
6	RRRRRR	88	9	5
7	LRLRLR	130	11	4
8	RLRLRL	135	9	8
9	LLR	69	4	12
10	LRLL	100	11	10
11	RLLR	120	9	6
12	RRLRL	150	16	8
13	RLRRRL	180	22	6
14	RRLRRR	220	15	7
15	LRLLRL	200	18	5

(the time of the information transmission does not depend on the text complexity) and H_1 (this time actually does depend on it).

In order to test H_1 against H_0, we carried out the following rank-order test. There are seven sequences of turns of equal length, 6, presented on lines 5–8 and 13–15 of Table 5.1. Let us order all these sequences based on the time ants spent transmitting them. Indexing the sequences by their line number in the table, we obtain 6,5,7,8,13,15,14. There are 7! possible orderings we could obtain, which are all equiprobable under H_0. Out of these orderings, only 2!2!3! are compatible with H_1. Indeed, as described above, there are three groups of 2, 2 and 3 strings such that strings in different groups have distinctly different complexities; these groups are: the sequences on lines (5, 6), lines (7, 8) and lines (13,14,15). The probability of obtaining one of the orderings compatible with H_1, under the assumption that H_0 holds, is (2!2!3!)/7! = 1/210, which is rather small. Since one of these orderings (the one given above) did indeed occur, H_0 should be rejected and H_1 accepted. Thus, the simpler the text the less time ants spent on transmitting it.

It is interesting that ants began to use regularities to compress only quite large sequences. Thus, they spent from 120 to 220 s to transmit information about random turn patterns on the maze with 5 and 6 forks and from 78 to 135 s when turn patterns were regular. On the other hand, there was no significant difference when the length of sequences was less than 4 (Table 5.1). These results show that ants are able to grasp regularities in a sequence of turns towards a target, to use these regularities to "compress" the sequence and finally to communicate this compressed information to others.

In sum, the use of the suggested information-theoretic approach helped us to reveal that the "language" of group–retrieving ant species exhibits such important properties as productivity, that is, the capacity for a potentially unlimited number of messages on the basis of finitely many signals, and flexibility, which allows signallers and perceivers to grasp regularities in their environment and use these regularities to optimise their communications.

References

Bradbury JW, Vehrencamp L (1998) Principles of animal communication. Sinauer Associates, Sunderland

Cilibrasi R, Vitányi PM (2005) Clustering by compression. IEEE Trans Inf Theory 51(4): 1523–1545

Kolmogorov AN (1965) Three approaches to the quantitative definition of information. Probl Inf Transm 1:1–7

Li M, Vitányi P (2009) An introduction to Kolmogorov complexity and its applications. Springer Science & Business Media, New York

McCowan B, Hanser SF, Doyle LR (1999) Quantitative tools for comparing animal communication systems: information theory applied to bottlenose dolphin whistle repertoires. Anim Behav 57(2):409–419

McCowan B, Doyle LR, Hanser SF (2002) Using information theory to assess the diversity, complexity, and development of communicative repertoires. J Comp Psychol 116(2):166

Reznikova Z, Ryabko B (1994) Experimental study of the ants communication system with the application of the Information Theory approach. Memorabilia Zoologica 48:219–236

Ryabko B, Reznikova Z (2009) The use of ideas of information theory for studying "language" and intelligence in ants. Entropy 11(4):836–853

Ryabko B, Astola J, Gammerman A (2006) Application of Kolmogorov complexity and universal codes to identity testing and nonparametric testing of serial independence for time series. Theor Comput Sci 359(1):440–448

Ryabko B, Reznikova Z, Druzyaka A, Panteleeva S (2013) Using ideas of Kolmogorov complexity for studying biological texts. Theory Comput Syst 52(1):133–147

Shannon CE (1948) A mathematical theory of communication. Bell Sys Tech J 1948(27):379–423, 623–656

Shannon CE (1951) Prediction and entropy of printed English. Bell Syst Tech J 30(1):50–64

Yaglom AM, Yaglom IM (1983) Probability and information. D. Reidel, Dordrecht

6 Ant "Language" Gives Insight into Studying Animal Numerical Competence

Basic number-related skills, that is, knowledge of quantities and their relations, are among the most exciting and intensively studied domains of animal behaviour. Perhaps, no field of ethological and cognitive science is based on comparison between animal and human abilities to such a great extent as the field of studying reasoning about numbers. What does numerical competence do with ants?

The use of ideas and methods of information theory has already helped us to reveal the fundamental features that ant "language" shares with human languages: the potentially unlimited (or at leas, great) number of messages, the correlation between the length of a message and the amount of information in it, and the similarity between ants' and humans' perceptions of complexity of a "text" (see Chap. 5). Further experiments described in this Chapter demonstrate that ants are able to encode and pass to each other information about numbers. These insects operate with numbers at a level that has not been demonstrated before in any non-human animals – not even in chimpanzees. However, we would not draw the conclusion that ants are champions of the Animal Kingdom in numerical competence. Rather, this means that the methods with which numerical skills in highly social animals have been studied so far have not been quite up to the task.

In this chapter, studies on ants' numerical abilities that are based on the information-theoretic approach are described. In particular, common rules in human and non-human reasoning about numbers are revealed. An experimental scheme for studying ants' arithmetic skills is based on a fundamental idea of information theory that in a "reasonable" communication system the frequency of usage of a message and its length must correlate.

Zh. Reznikova, *Studying Animal Languages Without Translation: An Insight from Ants*, DOI 10.1007/978-3-319-44918-0_6

6.1 Criteria of Numerical Competence for Comparative Studies

Approaches to studying numerical processing in animals are mainly based on criteria suggested by (Gelman and Gallistel 1978; Gallistel and Gelman 2000) for children and then adopted for animal studies by Davis and Pérusse (1988). There are four major processes involved in numerical competence: relative numerousness judgements, estimation, subitising and counting.

Until recently it was widely believed that the main difficulty in comparing numerical abilities in humans and other species is that number-related skills in our species are closely connected with our capacity for language and symbolic representation. However, recent studies revealed that not all human cultures possess linguistic means for number representation. The languages of several Amazonian indigenous groups lack words for numbers beyond 5, and these people fail in exact arithmetic with numbers larger than 4 or 5 (Pica et al. 2004; Dehaene et al. 2008; Everett and Madora 2012).

Development of cognitive ethology and behavioural ecology has given rise to a growing body of evidence that members of many species can judge about proportions and numbers of things, sounds, time intervals, smells, and so on. In nature, being able to perceive quantities is helpful in many situations, for example, in tracking predators, selecting the best foraging grounds or the best chance to mate. Different kinds of cognitive processes, including counting, can be understood in terms of the ways in which species are cognitively adapted to their different ecological niches. It is known that some species, both vertebrate and invertebrate, are capable of intellectual feats within narrow limits of solving certain vital problems. Humans have more difficulties in some tasks than animal "species genius", for example, in memorising locations (far lesser than food cashing birds and mammals) and in orienting in complex labyrinths (poorer than rats) (for a review, see Reznikova 2007).

Ants are good candidates for studying "species genius" in communication and number – related skills. There are many examples in the history of studying cognition in animals when the finding of a new, more adequate, experimental method has essentially changed the picture of the intellectual abilities of the animals under study, in particular of their numerical skills. Thus, applying a method of sample-choice to a chimpanzee revealed its ability to count within limits of two objects only (Ladygina-Kohts 2002/1935). The use of sequential responding task allowed researchers to show a subject of the same species to be able to respond adequately to Arabic numerals between 1 and 9 (Matsuzawa 1985). Further in this Chapter I consider how information theory approach helps us to reveal numerical competence in ants using their own communicative skills. Let us consider levels of numerical competence first, with application to animal cognition (for a detailed review, see: Reznikova and Ryabko 2011).

6.2 Experimental Studies of Numerical Skills in Animals

6.2.1 Relative Numerousness Judgements

In many experiments animals demonstrated that they perceive one stimulus as having "more" items compared with another that has "fewer". For example, in experiments with striped field mice *Apodemus agrarius* being presented with dangerous live prey (this time red wood ants served as victims, irrespective of their intelligence), when mice have had to choose between 5 vs. 15, 5 vs. 30, and 10 vs. 30 ants, they always tend to prefer the smaller quantity, thus displaying the capacity for distinguishing more from less in order to ensure comfortable hunting (Panteleeva et al. 2013).

Beetles (T*enebrio molitor*) were trained by using a spontaneous two-choice procedure in which males were exposed to substrates bearing odours from different numbers of females in increasing numerosity ratios (1:2, 1:3 and 1:4). Males discriminated sources of odours reflecting 1 versus 4 and 1 versus 3 females, but not 2 versus 4 or 1 versus 2 (Carazo et al. 2009). So, the beetles demonstrated the ability to discriminate between "more or less smells" within the limit of four with an operational signature ratio of 1:2. The similar results have been obtained on meadow voles (Ferkin et al. 2005; Ferkin 2015). Parental female cichlid spent more time trying to recover the fry from larger groups when tested with contrasts 6 versus 12 (1:2) and 6 versus 9 fry (2:3); however, they showed no preference in the 6 versus 8 (3:4) contrast, suggesting that this task exceeds their discrimination capacity (Forsatkar et al. 2016). In rhesus monkeys the ability to appreciate the abstract numerical equivalence between sets perceived in different sensory modalities such as tones and shapes has been demonstrated (Brannon and Merritt 2011).

It seems natural that, although accurate distinguishing between similar large quantities demands extensive training in animals, even after such training performance is much lower when compared to the performance on tasks involving only small quantities. For example, it is likely that primates could easily discriminate between a tree with 60 items on it and a one with 20. Moyer and Landauer (1967) first showed that when human adults are required to compare the relative magnitudes represented by two Arabic numerals, their reaction time is systematically influenced by both the linear distance and the absolute magnitude of the values compared. In other words, reaction time decreases with increasing numerical distance between two values (e.g., people are faster at 2 versus 9 than 2 versus 5 tasks), and if distance is held constant, reaction time increases with numerical magnitude (e.g., people are faster at 2 versus 3 than 4 versus 5 tasks). Experiments revealed similar regularities in rhesus monkeys (Cantlon and Brannon 2007) and even in newborn domestic chicks (Rugani et al. 2015), thus demonstrating similarities in the representation of magnitude in human- and non-human animals which does not demand language-related skills.

6.2.2 Subitising

Subitising is a form of pattern recognition that is used to rapidly assess small quantities of simultaneously presented items. The term has been coined by Kaufman et al. (1949) from the Latin word "subitus" meaning "sudden". Subitising can be described as an innate perceptual process through which a certain number of items can just be "seen" without the need for counting it.

To distinguish between subitising and other forms of numerical competence, the reaction time displayed by a subject is important. This criterion was originally derived from the reaction time displayed by humans in psychological experiments (Kaufman et al. 1949). This parameter was used in experiments of Matsuzawa (2003) with a female chimpanzee, Ai. She was trained to select an Arabic numeral corresponding to the number of objects within the quantity of 10. Ai showed quite an accurate and fast responding from 1 to 4. This was considered to be the same process as subitising in humans.

6.2.3 Estimation

Estimation in humans refers to the ability to assign a numerical label to a large array of items with a poor precision. When we judge at a glance that there are about 50 ducks on the lake we are "estimating".

This form of quantitative judgement have a great adaptive value in the context of natural life. For example, fish use number estimation in order to join to a greater shoal (Agrillo et al. 2008), female lions can judge about number of possible intruders by "counting" unfamiliar roars within the limit of four (McComb et al. 1994), New Zeland robins (*Petroica australis*) uses numerical judgements when retrieving and pilfering cashed food (Hunt et al. 2008), and ants of several species are able to estimate quite precisely numbers of encounters with members of other colonies on their feeding territories (Reznikova 1999; Gordon 2010).

Overall, the described experiments, conducted in a variety of species, demonstrate that animals enjoy different forms of quantitative judgements. Members of many species can compare two quantities when the differences between the quantities are big enough, and they are also capable of subitising, as well as of estimation of sets of stimuli.

6.2.4 Counting and Proto-Counting

Gelman and Gallistel (1978) list five criteria that formally define the process of counting and have been widely accepted in comparative studies. They are:

1. The one-to-one principle. Each item in a set (or event in a sequence) is given a unique tag, code or label so that there is a one-to-one correspondence between items and tags.

2. The stable-order principle (ordinality). The tags or labels must always be applied in the same order (e.g., 1, 2, 3, 4 and not 3, 2, 1, 4). This principle underlies the idea of ordinality: the label "3" stands for a numerosity greater than the quantity called "2" and less than the amount called "4".
3. The cardinal principle (cardinality). The label that is applied to the final item represents the absolute quantity of the set. In children, it seems likely that the cardinal principle presupposes the one-to-one principle and the stable-order principle and, therefore, should develop after the child has some experience in selecting distinct tags and applying those tags in a set.
4. The abstraction principle (property indifference). Counting can be applied to heterogeneous items. In experiments with children, a child should be able to count such different items as toys of different kinds, colour or shape and to demonstrate skills of counting even actions or sounds. There are indications that many 2 or 3 year old children can count mixed sets of objects.
5. The order irrelevance: the order in which objects are counted is irrelevant.

Most researchers define "true counting" as a formal enumerative process that conforms to the principles proposed by Gelman and Gallistel (1978), and use the term "proto-counting" as a more flexible one, thus leaving open the question of whether or not animals can count in the same sense that humans can count. However, there is some evidence of similarities between processes of counting in human- and non-human animals. For example, chimpanzees, like young children, tend to touch or point to each item when judging the number of items in an array (Boysen et al. 1996). These gestures, known as indicating acts, may help the child or animal coordinate the tagging process involved in the application of the one-to-one correspondence principle.

There is a great deal of experimental results supporting the hypotheses that members of many species can recognize the ordinal character of numbers (Boysen 1993; Xia et al. 2000; Pepperberg 2006; Rugani et al. 2015). For example, capuchin monkeys and rhesus monkeys trained to associate Arabic numerals with the corresponding number of food pellets were able to order correctly arrays of up to five numerals. This behaviour suggests that they understand the order of the symbols, as required by the stable order principle. Chimpanzees can reliably apply the correct Arabic numeral to arrays of familiar and novel objects. They can also select the number of dots on a computer screen equal to an Arabic numeral cue and then indicate the end of the count by contacting the numeral. This behaviour further demonstrates the use of the cardinality principle (for a review, see Beran et al. 2015).

Experiments of Chittka and Geiger (1995) suggested that honey bees can count landmarks or at least use the number of landmarks as one of the criteria in searching for food sources. Researchers worked with individual honey bees in medow consisted of large yellow tents. The bees were trained to take sugar syrup from a feeder that was placed between the third and fourth tents. In the tests, the number of landmarks between hive and feeder were altered. It is interesting to note that individual foragers in a hive used different cues in their searching. Many bees continued to rely only on flying distance between the hive and the feeder. Anyway, the distance estimation of the bees as a group depended notably on the number of landmarks. If some family members encountered more landmarks on their way from the hive to the feeder than they had during training, they landed at a shorter distance than during control tests with the training landmark set up. If they encountered fewer landmarks, they flew significantly further.

Discussing their data, the authors consider it unlikely that their bees meet the abstraction principle of "true counting", and they referred the observed behaviour as proto-counting.

Dacke and Srinivasan (2008) obtained further results on counting in honey bees training them to forage from a tunnel. Bees received a food reward after they have passed a specific number of landmarks. They appeared to be able to count up to four objects when these objects were encountered sequentially during flight. Bees trained in this way were able to count novel objects which they have never previously encountered, thus demonstrating that they are capable of object-independent counting, again, within a limit of four.

Capaldi and Miller (1988) tested rats' ability to detect absolute number by using biologically important events – the number of reinforced runs followed by a non-reinforced run – and found that the rats ran significantly slower on the non-reinforced run, and they distinguish between "numbers" of runs. Rayburn-Reeves et al. (2010) used a similar procedure in their experiments with pigeons. The birds were given a sequence of trials in which responding on the first three trials ended in reinforcement but responding on the fourth trial did not (RRRN). When the response requirement on each trial was a single peck, no significant increase in latency to peck on the fourth trial was found. When the response requirement was increased to 10 pecks, however, the time to complete the peck requirement was significantly longer on the non-reinforced trial than on the reinforced trials. Tests for control by time, number of responses, and amount of food consumed indicated that the pigeons were using primarily the number of reinforcements obtained in each sequence as a cue for non-reinforcement.

In general, the five criteria listed by Gelman and Gallistel (1978) were used in the experiments on counting in animals only as a very general guideline, with some of the principles verified for some species and others for others (perhaps, all the five principles were established only for chimpanzees that have used Arabic numerals). Therefore, for the sake of clarity of definitions, and to comply with the established terminology, we can say that in these experiments (excluding some of those with chimpanzees) it was only proto-counting that was demonstrated.

6.2.5 Arithmetic Skills

One of the most interesting fields in cognitive ethology is studying arithmetic abilities in non-human animals and human infants. Wynn (1992) explored whether 5-month old infants can solve addition and subtraction problems. She used "looking time" (the expectancy violation technique) as a relevant measure to judge about subject's understanding of a problem. The idea was that if infants keep track of the numbers of toys they see being placed behind a screen, they should look longer at a screen that, when lowered, reveals an outcome that violates their expectations. For example, in the "$(1+1)=1$ or 2" task, one doll was placed on a stage, covered with a screen, and then another doll was visibly introduced behind the screen. A further manipulation occurred out of sight of the subjects, where either another doll was added, or one of the existing ones was removed. When the screen was removed, infants looked longer at the "impossible" outcomes of either one doll or three dolls, suggesting that they expected two dolls. The same kind of result emerges from an experiment involving subtraction instead of addition. Wynn concluded that infants have an innate capacity to do simple arithmetic.

Hauser et al. (2000) set up the same task for the free-living rhesus monkeys. In the first series of experiments subjects watched as an experimenter placed two eggplants behind a screen and then removed the screen. Subjects looked longer when the test outcome was one or three eggplants than when it was the expected two. Like human infants, rhesus monkeys appear to understand that $1+1=2$. Rhesus monkeys also appear to understand that $2+1=3$, $2-1=1$, and $3-1=2$. It is worth noting that this technique allows to reveal what animals "think" spontaneously, in the absence of training. Comparable results have been obtained with cotton-top tamarins (Uller et al. 2001) and lemurs (Santos et al. 2005).

Using other experimental schemes, researchers have demonstrated animals' abilities for mental manipulations with quantities. Summation has been demonstrated in tasks where animals were required to compute and choose the larger of two quantities. For example, Boysen and Berntson (1989) provided an impressive demonstration of arithmetic abilities in non-humans. A chimpanzee Sheba, experienced with Arabic numerals, was led around a room to three different hiding places that could contain 1, 2, 3, or 4 pieces of orange. Subsequently, Sheba was required to choose the Arabic numeral that corresponded to the sum of number of items she had met before. She chose the correct sum from the first session of test in both cases, showing mastery of this ability without any explicit training.

Cantlon and Brannon (2007) tested two rhesus monkeys and college students in a task where two sets of dots were shown in succession on a computer screen, and participants had to add the sets and then find a matching option that had the same total number of dots. For example, subjects could have been shown two dots and four dots, followed by a choice of six dots —the right answer – or, say, eight dots – a wrong answer. While the college students were correct 94 % of the time and the monkeys 76 %, the average reaction time for both monkeys and humans was about 1 s. The performance of both the monkeys and the college students worsened when the two choices were closer.

There are also some studies on arithmetic skills beyond primates. For example, in experiments with pigeons Brannon et al. (2001) found that they are able to discern correct from incorrect outcomes of a subtraction task. Pigeons were required to compare a constant number with the number remaining after a numerical subtraction. The birds appeared to be able to subtract one set of light flashes from another and, subsequently, peck the estimated number of flashes that are left after subtraction. Pigeons successfully solved different tasks including "12−6". The language-trained grey parrot Alex, able to quantify sets of 6 or fewer items by using English labels, was tested on tasks involving addition of quantities up to 6. He was, without explicit training, asked "How many total X?" for two sequentially presented collections and required to answer with a vocal English number label. His accuracy suggested that his addition abilities are comparable to those of non-human primates and young children (Pepperberg 2006).

Experiments on newly hatched domestic chicks (Rugani et al. 2010) enable researchers to appreciate core components of animals' numerical cognition, that is, a set of building block systems that emerge early in ontogeny and show characteristic limits of domain and task specificity. Chicks were reared with five identical objects (small balls). On days 3 or 4, chicks underwent free-choice tests in which two sets containing three and two balls disappeared (either simultaneously or one by one), each behind one of two opaque identical screens. Chicks spontaneously inspected the screen occluding the larger set. In the next experiment, after the initial disappearance of the two sets, some of the objects were visibly transferred, one by one, from one screen to the other. Computation of a series of subsequent additions or subtractions of elements that appeared and disappeared, one by one, was needed in order to perform the task successfully. Chicks chose the screen, hiding the larger number of elements at the end of the event, irrespective of the directional cues provided by the initial and final displacements.

Thus, not only primates but some birds including newly hatched domestic chicks displayed the ability to add and subtract small numbers. In this context, it is worth of noting that demonstrating the animals' ability to simple arithmetic does not require demonstrating (as a prerequisite) their ability to count.

Overall, although many impressive results concerning sophisticated forms of numerical cognition in animals have been obtained, our knowledge is restricted by the use of artificial communicative systems elaborated specifically for communication with animals, or by the use of experimental paradigms for examining individual cognitive capacities in rather sophisticated situations of solving complex learning problems. In this respect, we encounter a methodological paradox. It is evident that in humans high levels of number-related skills are closely connected with the development of language. At the same time, all known experimental paradigms for studying numerical competence in animals do not exploit the phenomenon of close relations between intelligence, sociality and natural communication. Even in the honey bee studies mentioned above (Chittka and Geiger 1995; Dacke and Srinivasan 2008), cognitive capacities of these social insects have been tested individually, and the facilities of honey bees' extraordinary symbolic language were not included into the experimental schemes. In Sect. 6.3 we consider a principally different approach for studying numerical competence in animals, which harnesses their natural communication.

6.3 An Insight from Ant "Language": Flexibility of Communication as a Basis for Arithmetic Skills

Our studies on numerical competence in ants are based on the fact that they are able to pass exact information about a target in a maze, revealed earlier in the binary tree study (Chap. 5). In our "counting ant" study (Reznikova and Ryabko 1994, 2009, 2011, 2012) scouting individuals were required to transfer to foragers in a laboratory nest the information about which branch of a special "counting maze" they had to go to in order to obtain syrup.

"Counting maze" is a collective name for several variants of set-ups containing different "branches" with troughs. The experiments were based on a procedure similar to the binary tree study. The main idea of this experimental paradigm is that experimenters can judge how ants represent numbers by estimating how much time individual ants spend on "pronouncing" numbers, that is, on transferring information about index numbers of branches.

The first variant of the counting maze is a comb-like set-up consisting of a long horizontal plastic trunk with 25–60 equally spaced plain plastic branches, each of them 6 cm in length (Figs. 6.1 and 6.2). Each branch ended with an empty trough, except for one filled with syrup. Ants came to the initial point of the trunk over a

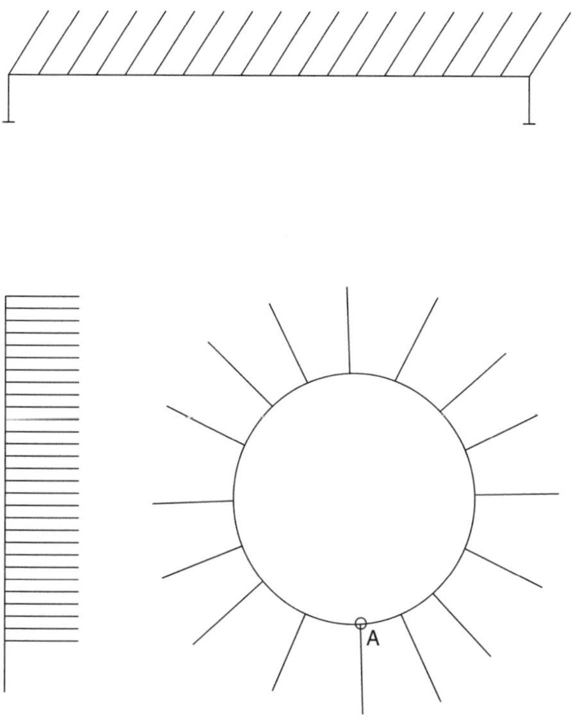

Fig. 6.1 "Counting mazes": a *horizontal trunk*, a *vertical trunk* and a *circle*

Fig. 6.2 An ant team on the horizontal trunk: no food this time (Photo by Zhanna Reznikova)

small bridge. The second variant is a set-up with 60 vertically aligned branches. In order to test whether the time of transmission of information about the number of a branch depends on its length as well as on the distance between the branches, one set of experiments was carried out on a similar vertical trunk in which the distance between the branches was twice as large, and the branches themselves were three times and five times longer (for different series of trials). The third variant was a circular trunk with 25-cm-long branches.

Similarly to the binary tree study (Chaps. 4 and 5), ants were housed in a laboratory arena divided into two parts, one containing a plastic nest with a laboratory ant colony and the other containing one of the variants of the counting maze. All actively foraging ants were individually marked with coloured paint. Each series of experiments was preceded by the run-up stage consisting of familiarization trials. In order to force a scout to transfer the information about food to its nestmates the experimenters showed it the trough containing syrup (placing the scout directly on the trough) and then let it return to the nest. Again, similarly with the binary tree study, the time, which the scouting ants spent on the "informative contacts" with their teams (as it was described in Sect. 5.2) was measured (in seconds). The duration of each information contact was considered to begin when the scout touched the first forager ant, and to end when the first two foragers left the nest for the maze. After allowing it to contact the foragers within the nest, the scout was removed and isolated for a while, so that the foragers had to search for the food by themselves, without their guide.

As it was described in Sects. 4.1 and 5.2, the experiments were devised so as to eliminate all possible ways for the members of each foraging team to find the target, except by an information contact with their scout. The set-up was replaced with a fresh one, with all troughs filled with water, while the scout was in the nest; if the foraging team reached the correct branch in a compact group, then the water-filled trough was replaced with a one with syrup. It is worth to repeat that in these series of experiments ants had to face a rather complex life-or-death task: they could obtain food only in the maze and only once every 2–3 days.

6.3.1 The Ants' Ability to Transfer Information About Numbers of Objects

The findings described below concerning number-related skills in ants are based on comparisons of duration of information contacts between scouts and foragers that preceded successful trips by the foraging teams. Two laboratory colonies of red wood ants *F. polyctena* were used in this set of experiments. In total, 32 scout-foragers teams worked in three kinds of set-ups. The teams left the nest after they were contacted by scouts and moved towards the trough by themselves on 152 trials. In 117 cases the team immediately found the correct path to the trough. In the remaining cases, ants came to the empty troughs, and began looking for food by checking neighbouring branches.

It was proofed statistically that the success ratio which was obtained experimentally could only be explained by information transmission from the scouts. In addition, in control experiments ants, including scouts, placed in the set-up, without information on which trough contained food usually failed to find the food, even though they actively searched for it (see details in: Ryabko and Reznikova 2009; Reznikova and Ryabko 2011).

Data obtained on the vertical trunk are shown in Table 6.1 as an example. It turned out that, when the rewarded branch was randomly selected by the experimenters, the relation between the number of the branch (i) and the duration of the contact between the scout and the foragers (t) was well described by the equation $t = ai + b$ (where a and b are constant) for different set-ups which were characterised by different shapes, distances between the branches and lengths of the branches.

Table 6.1 The results of experiments in the "vertical trunk 1" with *F. polyctena*

Exp. Nr.	Number of food-containing branch	Duration of scout-forager contact (s)	Working team number (or, what is the same, the scout's "name")
1	10	42	I
2	10	40	II
3	10	45	III
4	40	300	II
5	40	280	IX
6	13	90	II
7	13	98	I
8	28	110	III
9	28	120	X
10	20	120	X
11	20	110	III
12	35	260	III
13	35	250	X
14	30	160	I
15	30	170	III

Table 6.2 Values of correlation coefficient (r) and regression (a, b) coefficients for vertical trunk (vert), horizontal trunk (horiz), and circle in the experiments with *F. polyctena*

Type of setup	Sample size	Number of branches	r	a ± Δ a	b ± Δ b
Vert.1	15	40	0.93	7.3 ± 4.1	−28.9 ± 0.51
Vert.2	16	60	0.99	5.88 ± 0.44	−17.11 ± 0.65
Horiz.1	30	25	0.91	8.54 ± 1.1	−22.2 ± 0.62
Horiz.2	21	25	0.88	4.92 ± 1.27	−18.94 ± 4.75
Circle	38	25	0.98	8.62 ± 0.52	−24.4 ± 0.61

The parameter *b* is introduced, since ants can transmit information not related directly to the task, for example, the additional signal "food". Besides, it is not ruled out that a scouting ant transmits, in some way, the information on its route to the nest, using acoustic or some other means of communication. In this context, it is important that the route from the maze to the nest on the arena was in all experiments approximately the same. Being highly motivated, scouts hurried on to the nest in a beeline and, therefore, the time before they made antennal contact with the foragers in the nest, which the scout could hypothetically use for message transmission, was approximately the same and did not depend either on the lengths of the branches or on other features of set-ups. The values of parameters *a* and *b* appeared to be close for all set-ups, and did not depend on their physical characteristics. The correlation coefficient between *t* and i was high for different kinds of counting mazes. In turn, it proves the hypothesis that that the connection between *t* and *i* is close to linear (Table 6.2).

The likely explanation of the results concerning ants' ability to find the "right" branch is that they can evaluate the number of the branch in the sequence of branches in the maze and transmit this information to each other. Presumably, a scout could pass messages not about the number of the branch but about the distance to it or about the number of steps and so on. What is important is that even if ants operate with distance or with the number of steps, this shows that they are able to use quantitative values and pass on exact information about them.

It is worth to note that in other series of experiments made by a similar procedure with five colonies of *F. polyctena*, these ants were demonstrated as being able to transmit information on the ordinal number of a given side-branch in a linear maze (a "lattice" set-up) and on the numerical coordinates in a grid-type maze of a node (a "globe" set-up) (Fig. 6.3). The results of these experiments showed that the ants are able to memorise and transmit information regarding the location of the target on the plane or on the globe surface (Reznikova and Ryabko 2001).

All these data based on the analysis of the time spent by ants for transmitting the coordinates of objects in different situations enable us to conclude that these insects can encode and transmit information about quantitative parameters of objects.

As in the experiments with other non-human species (described in Sect. 6.3), in ants it was only proto-counting that was established, since adherence to all of the five principles of counting listed by Gelman and Gallistel (1978) was not demonstrated.

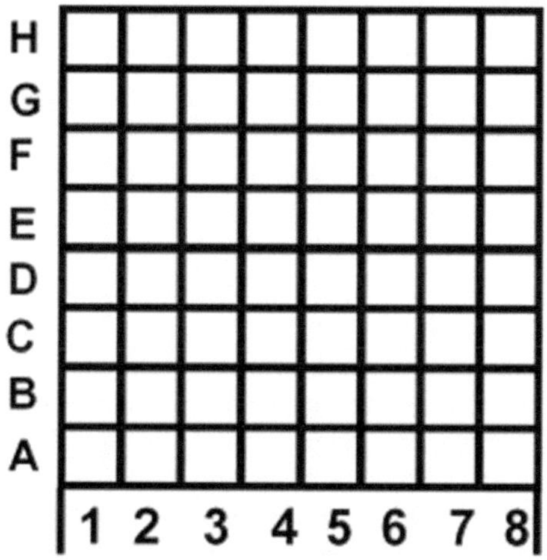

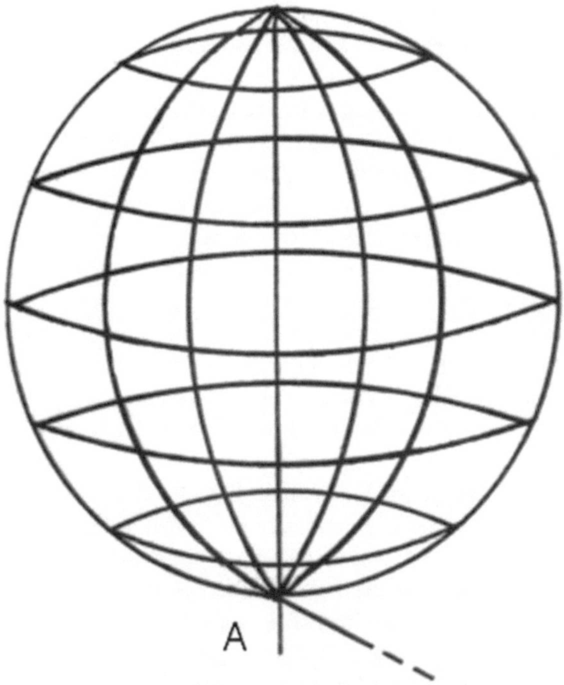

Fig. 6.3 A "lattice" set – up and a "globe" set – up

6.3.2 The Ants' Ability to Add and Subtract Small Numbers

The experimental paradigm of studying ants' "arithmetic" skills is based on a fundamental idea of information theory, which is that in a "reasonable" communication system the frequency of usage of a message and its length must correlate. This rule is described by the Shannon's equation connecting the length of a message (l) and its frequency (p), i.e., $l = -\log p$. The informal pattern is quite simple: the more frequently a message is used in a language, the shorter is the word or the phrase coding it. Professional slang, abbreviations, etc. can serve as examples. This phenomenon is manifested in all known human languages as well as in technical systems of information transmission.

The experimental procedure was similar to other experiments with counting mazes (Reznikova and Ryabko 2011). In various years four colonies of red wood ants *F. polyctena* were used. The scheme of the experiments is as follows. Ants were offered a horizontal trunk with 30 branches. Each experiment was divided into three stages, and at each of them the regularity of placing the trough with syrup on branches with different numbers was changed. At the first stage, similarly to the previous set of experiments, the branch containing the trough with syrup was selected randomly, with equal probabilities for all branches. So the probability of the trough with syrup being placed on a particular branch was 1/30. Thus, the first stage of the "arithmetic" experiments did not differ from "counting" experiments described above. At the second stage two "special" branches A and B were chosen (N 7 and N 14; N 10 and N 20; and N 10 and N 19 in different years) on which the trough with syrup was placed much more frequently than on the rest of the branches: with the probability 1/3 for A and B, and 1/84 for each of the other 28 branches. In this way, two "messages" to be transmitted, namely, "the trough is on branch A" and "the trough is on branch B", had a much higher probability than the remaining 28 messages. In one series of trials we used only one "special" point A (the branch N 15). On this branch the food appeared with the probability of 1/2, and 1/58 for each of the other 29 branches. At the third stage of experiments the number of the branch with the trough was again chosen at random.

Now let us consider the relationship between the time which the ants spent to transmit the information about the branch containing food, and its number. The information obtained at the first and third stages of the experiments are shown on the graph (Fig. 6.4) in which the time of the scout's contact with foragers (t) is plotted against the number (i) of the branch with the trough. At the first stage the dependence is close to linear. At the third stage, the picture was different: first, the information transmission time was very much reduced, and, second, the dependence of the information transmission time on the branch number is obviously non-linear: a depression can be seen in the vicinities of the "special" points (A and B). So the experimental data demonstrate that the patterns of the dependence between the time of information transmission and the index number of the food-containing branch at the first and third stages of experiments are considerably different. Moreover, in the vicinities of the "special" branches, the time taken for transmission of the information about the number of the branch with the trough is, on the average, much shorter.

For example, in the first series, at the first stage of the experiments the ants took 70–82 s to transmit the information about the fact that the trough with syrup was on branch N 11, and 8–12 s to transmit the information about branch N 1. At the third stage it took 5–15 s to transmit the information about branch N 11 (nearest to the "special" branch N 10). These data enable us to suggest that the ants have changed the mode of presenting the data about the number of the branch containing food.

The statistical analysis of the results (see: Ryabko and Reznikova 2009; Reznikova and Ryabko 2011) supports the hypothesis that at the third stage of the experiment

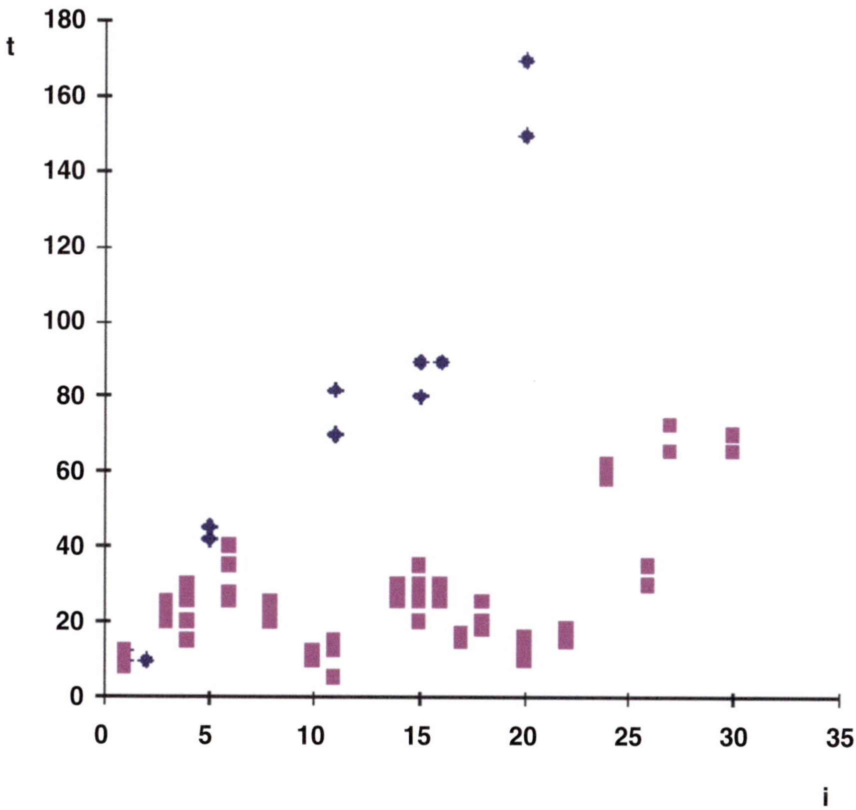

Fig. 6.4 Dependence of the time (in seconds) of transmission of information about the number of the branch having food (*t*) on its ordinal number (*i*) at the first and the third series of experiments in the ant *Formica polyctena*. Diamonds, the time taken for transmission of information at the first stage; *Squares*, the same, at the third stage

the time of transmission of a message about the number of the branch is shorter when this branch is close to either of the special ones. Our interpretation is that at this stage of the experiment the ants used simple additions and subtractions, achieving economy in a manner reminiscent of the human numerical system. Note that when using numerical systems, people have to perform simple arithmetical operations, for example, 13 = 10 + 3. It is particularly obvious in Roman numerals, for example, VII = V + II.

Let us now consider ants' arithmetic skills in more details. Analysis of the time duration of information transmission by the ants raises the possibility that at the third stage of the experiments the scouts' messages consisted of two parts: the information about which of the "special" branches was the nearest to the branch with the trough, and the information about how many branches away is the branch with the trough from a certain "special" branch. Informally, the ants were forced to develop a new code based on simple arithmetic operations, that is, to perform an operation similar to passing the "name" of the "special" branch nearest to the branch with the trough, followed by the number which had to be added or subtracted in order to find the branch with the trough. A good illustration here is that ants spend, for example, 25–30 s to transmit the message "a trough is on the branch 4" at the first stage of the experiment, and they spend the same time to transmit a message that says "a trough is on the forth branch after the branch 10" at the third stage. That ant teams went directly to the "correct" branch enables us to conclude that they performed correctly whatever "mental" operation was to be made.

Let us show that the mental operations that the scouts had to perform can be considered the operations of addition and subtraction. Indeed, we have shown that what the scout had to do was to transmit the name of one of the "special" branches plus the directions (how many branches backwards or forwards) from this special branch. However, in these experiments the "special" branches could only be referred to by their numbers. That is, there was no possibility to associate some visual (olfactory, etc.) mark with them.

In the absence of any such mark, what scout had to do when placed on, say, the branch 17, when the branch 20 was "special", was to calculate correctly that 17 is 3 branches away from 20 (the "special" one). In other words, it had to perform subtraction. Similarly, when placed on the branch (say) 23, the scout had to correctly compute that this branch is 3 branches away from the "special" branch 20, which, in the absence of any mark on the branch 20, requires addition.

The foragers then had to correctly interpret this information, which does not necessarily mean that they have actually performed exactly the same mental operation. For example, if the information from the scout was of the form "20 + 3", then the foragers could, when passing the branch 20, start "counting" to 3 to reach the target. This does not require performing the operation of addition.

In order to verify this statistically, the coefficient of correlation was calculated between the time required for transmission of information about the trough being on the branch i and the distance from i to the nearest "special" branch. The results confirmed the hypothesis that the time for transmission of a message about the number of the branch is shorter when this branch is closer to any of the "special" ones. For this purpose, the data obtained at the third stage of the experiment were transformed to present them in the form shown in Table 6.3 where data of 1 year are given as an example. In this table we do not include branches that are close to the starting point of the set-up (N 1–4) because there is no need to use "arithmetic" for ants where rewarded branches are very close to the first one (in fact ants spent roughly the same time transmitting information about these branches: from 10 to 20 s).

It can be seen from Table 6.4 that the coefficients of correlation between the transmission time and the distance to the nearest special point have quite high values and they differ significantly from zero. So the results support the hypothesis that the time for transmission of a message about the number of the branch is shorter when this branch is close to either of the special ones. This, in turn, shows that at the third stage of the experiment the ants used simple additions and subtractions, achieving economy in a manner reminiscent of the Roman numeral system when the numbers 10 and 20, 10 and 19 in different series of the experiments, played a role similar to that of the Roman numbers V and X.

The obtained results also demonstrate that red wood ants possess a communication system that has a great degree of flexibility. Until the frequencies with which

Table 6.3 Dependence of the time of information transmission (t) on the distance from the branch with a trough to the nearest "special" branch (special branches are 10 and 20)

The number of the branch having food (experiments in different days, consequently)	Distance to the nearest "special" branch	Times of transmission of information about the branch number for different scouts (sec)
26	6	35, 30
30	10	70, 65
27	7	65, 72
24	4	58, 60, 62
8	2	22, 20, 25
16	4	25, 8, 25
16	4	25
22	2	15, 18
18	2	20, 25, 18, 20
15	5	30, 28, 35, 30
20	0	10, 12, 10
6	4	25, 28
16	4	30, 25
15	5	20, 25, 20
14	4	25, 28, 30, 26
17	3	17, 15
11	1	10, 12

Table 6.4 Values of correlation coefficient (r) in the experiments with different "special" branches

Sample size	Numbers of "special" branches	r for the first stage of the experiments	r for the third stage of the experiments
150	10, 20	0.95	0.80
92	10, 19	0.96	0.91
99	15	0.99	0.82

the food was placed on different branches started exhibiting regularities, the ants were "encoding" each number i of a branch with a message of length proportional to i. Subsequent changes of the code in response to special regularities in the frequencies are in line with one of the basic information-theoretic principles that states that in an efficient communication system the frequency of use of a message and the length of the message are related. The complexity and flexibility of ants' communication can be considered an evidence of their intelligence.

Apparently, the fact that different forms and elements of numerical competence are spread across a wide variety of species, both vertebrate and invertebrate, as well as that these abilities emerge early in ontogeny, support the idea of evolutionary psychology inspired by Darwin (1871) that the human mind is a collection of special-purpose mechanisms, each shaped by evolution to perform a particular function. A huge body of experimental data includes demonstrations of animals' abilities to count, to understand numerical information and to perform simple arithmetic operations. Not only primates but also ants, honey bees, pigeons, and even newly hatched domestic chicks demonstrate some forms of numerical competence. More simple (but also cognitively demanding) number – related skills, such as relative numerousness judgements, estimation and subitising have been revealed in a variety of species, from beetles, fish and salamanders to rodents, dogs, cats, horses, elephants and primates (for a review, see Reznikova and Ryabko 2011).

There is much work to be done to extend our understanding of whether at least some species share advanced characteristics of numerical cognition with humans, or whether animals think about the quantitative parameters of the world in a way radically different from ours. Although our knowledge about advanced forms of animal numerical competence have been much enlarged by applying new experimental paradigms and elaborating new concepts, we are still far from having an integrative measure of animal numerical competence. New approaches are needed, and in this aspect the insight from ants is indicative.

The information-theoretic experimental paradigm described in this Chapter has been used to show that members of highly social ant species are able to pass information about numbers of objects (at least within 30), and, moreover, they can add and subtract small numbers and use these operations for optimising their messages. As already mentioned, the scheme of the experiments is based on the information-theoretic idea that in a "reasonable" communication system the frequency of usage of a message and its length must correlate. Facing a new task in which a food source appeared on a "special" branch of a counting maze much more frequently than on other ones, scouting ants were forced to develop a new code based on arithmetic operations.

The obtained results do not mean that within the animal kingdom ants possess outstanding cognitive skills. First, it is likely that these cognitive abilities lie within a narrow domain of ants "species genius" and concern their complex and flexible communication system which has much in common with the honey bee "dance language" and serves for transferring information about food. Second, which is particularly important, the described situation is, as far as we know, the first case when animals' own communicative means are used for studying their numerical cognition. The experimental paradigm based on the information-theoretic approach does not require the subjects to solve any artificial learning problems, such as learning intermediary languages, or even learning to solve multiple-choice problems. The main principle of this scheme of experiments is to force animals to transfer a piece of information known exactly to the researchers, and to measure and analyse the time of information transmission.

It can be suggested that the scheme of experiments used to study ants' numerical competence can be directly used on other animals, including mammals, and, in particular those whose intellectual abilities are most studied – the chimpanzees. Using this scheme on such animals as chimpanzees would require the use of cooperative tasks. For example, two subjects can be involved, one of which is shown the food source, but has to communicate its location to the other, in order for both of them to get it. The food can be placed in one out of (say) 20 containers, which play the same role as the branches of the comb-like maze in the experiments with ants. The numerical abilities of the subjects can then be studied by analysing the time the first subject (that has seen the food) spends on transferring the information on its location to the other, exactly as in the experiments with ants.

Therefore, we suggest that the use of communication systems of highly social animals and application of ideas of information theory can open new horizons for studying numerical competence in non-human animals.

References

Agrillo C, Dadda M, Serena G, Bisazza A (2008) Do fish count? Spontaneous discrimination of quantity in female mosquitofish. Anim Cogn 11(3):495–503

Beran MJ, Perdue BM, Evans TA (2015) Monkey mathematical abilities. In: Cohen Kadosh R, Dowker A (eds) Oxford handbook of mathematical cognition. Oxford, UK: Oxford University Press, p 237

Boysen ST (1993) Counting in chimpanzees: nonhuman principles and emergent properties of number. In: Boysen ST, Capaldi EJ (eds) The development of numerical competence: animal and human models. Lawrence Erlbaum Associates, Hillsdale, pp 39–59

Boysen ST, Berntson GG (1989) Numerical competence in a chimpanzee (Pan troglodytes). J Comp Psychol 103(1):23

Boysen ST, Berntson GG, Hannan MB, Cacioppo JT (1996) Quantity-based interference and symbolic representation in chimpanzees (*Pan troglodytes*). J Exp Psychol Anim Behav Process 22(1):76–86

Brannon EM, Merritt D (2011) Evolutionary foundations of the approximate number system. In: Space, time, and number in the brain: searching for the foundations of mathematical thought, New York, NY: Elsevier p 207–224

Brannon EM, Wusthoff CJ, Gallistel CR, Gibbon J (2001) Numerical subtraction in the pigeon: evidence for a linear subjective number scale. Psychol Sci 12(3):238–243

Cantlon JF, Brannon EM (2007) How much does number matter to a monkey (Macaca mulatta)? J Exp Psychol Anim Behav Process 33(1):32

Capaldi EJ, Miller DJ (1988) Counting in rats: its functional significance and the independent cognitive processes that constitute it. J Exp Psychol Anim Behav Process 14(1):3

Carazo P, Font E, Forteza-Behrendt E, Desfilis E (2009) Quantity discrimination in Tenebrio molitor: evidence of numerosity discrimination in an invertebrate? Anim Cogn 12(3):463–470

Chittka L, Geiger K (1995) Can honey bees count landmarks? Anim Behav 49(1):159–164

Dacke M, Srinivasan MV (2008) Evidence for counting in insects. Anim Cogn 11(4):683–689

Darwin C (1871) The descent of man, and selection in relation to sex, Vol. 1, 2 John Murray London

Davis H, Pérusse R (1988) Numerical competence in animals: definitional issues, current evidence, and a new research agenda. Behav Brain Sci 11(04):561–579

Dehaene S, Izard V, Spelke E, Pica P (2008) Log or linear? Distinct intuitions of the number scale in Western and Amazonian indigene cultures. Science 320(5880):1217–1220

Everett C, Madora K (2012) Quantity recognition among speakers of an anumeric language. Cognit Sci 36(1):130–141

Ferkin MH (2015) The response of rodents to scent marks: four broad hypotheses. Horm Behav 68:43–52

Ferkin MH, Pierce AA, Sealand RO (2005) Meadow voles, *Microtus pennsylvanicus*, can distinguish more over-marks from fewer over-marks. Anim Cogn 8(3):182–189

Forsatkar MN, Nematollahi,MA, Bisazza A (2016). Quantity discrimination in parental fish: female convict cichlid discriminate fry shoals of different sizes. Anim Cogn 19(5):959–964

Gallistel CR, Gelman R (2000) Non-verbal numerical cognition: from reals to integers. Trends Cogn Sci 4(2):59–65

Gelman R, Gallistel CR (1978) The child's understanding of number. Cambridge MA Harvard University Press

Gordon DM (2010) Ant encounters: interaction networks and colony behavior. Princeton University Press, Princeton

Hauser MD, Carey S, Hauser LB (2000) Spontaneous number representation in semi–free–ranging rhesus monkeys. Proc R Soc Lond B Biol Sci 267(1445):829–833

Hunt S, Low J, Burns KC (2008) Adaptive numerical competency in a food-hoarding songbird. Proc R Soc Lond B Biol Sci 275(1649):2373–2379

Kaufman EL, Lord MW, Reese TW, Volkmann J (1949) The discrimination of visual number. Am J Psychol 62(4):498–525

Ladygina-Kohts NN (2002/1935) Infant chimpanzee and human child. Oxford University Press, Oxford

Matsuzawa T (1985) Use of numbers by a chimpanzee. Nature 315:57–59

Matsuzawa T (2003) The Ai project: historical and ecological contexts. Anim Cogn 6(4):199–211

McComb K, Packer C, Pusey A (1994) Roaring and numerical assessment in contests between groups of female lions, *Panthera leo*. Anim Behav 47(2):379–387

Moyer RS, Landauer TK (1967) The time required for judgments of numerical inequality. Nature 215:1519–1520

Panteleeva S, Reznikova Z, Vygonyailova O (2013) Quantity judgments in the context of risk/reward decision making in striped field mice: first "count", then hunt. Front Psychol 4:53 (Number without language: comparative psychology and the evolution of numerical cognition, 45)

Pepperberg IM (2006) Grey parrot (*Psittacus erithacus*) numerical abilities: addition and further experiments on a zero-like concept. J Comp Psychol 120(1):1

Pica P, Lemer C, Izard V, Dehaene S (2004) Exact and approximate arithmetic in an Amazonian indigene group. Science 306(5695):499–503

Rayburn-Reeves RM, Miller HC, Zentall TR (2010) "Counting" by pigeons: discrimination of the number of biologically relevant sequential events. Learn Behav 38(2):169–176

Reznikova Z (1999) Ethological mechanisms of population density control in coadaptive complexes of ants. Russ J Ecol 30(3):187–192

Reznikova Z (2007) Animal intelligence: from individual to social cognition. Cambridge University Press, Cambridge

Reznikova Z, Ryabko B (1994) Experimental study of the ants communication system with the application of the Information Theory approach. Memorabilia Zoologica 48:219–236

Reznikova Z, Ryabko B (2001) A study of ants' numerical competence. Electron Trans Artif Intell B 5:111–126

Reznikova Z, Ryabko B (2011) Numerical competence in animals, with an insight from ants. Behaviour 148(4):405

Reznikova Z, Ryabko B (2012) Ants and bits. Plenary talk presented at the 2011 IEEE International Symposium of Information Theory. IEEE Inf Theory Soc Newsl 62(5):17–20

Rugani R, Regolin L, Vallortigara G (2010) Imprinted numbers: newborn chicks' sensitivity to number vs. continuous extent of objects they have been reared with. Dev Sci 13(5):790–797

Rugani R, Vallortigara G, Priftis K, Regolin L (2015) Number-space mapping in the newborn chick resembles humans' mental number line. Science 347(6221):534–536

Ryabko B, Reznikova Z (2009) The use of ideas of information theory for studying "language" and intelligence in ants. Entropy 11(4):836–853

Santos LR, Barnes JL, Mahajan N (2005) Expectations about numerical events in four lemur species (*Eulemur fulvus*, *Eulemur mongoz*, *Lemur catta* and *Varecia rubra*). Anim Cogn 8(4):253–262

Uller C, Hauser M, Carey S (2001) Spontaneous representation of number in cotton-top tamarins (*Saguinus oedipus*). J Comp Psychol 115(3):248

Wynn K (1992) Addition and subtraction by human infants. Nature 358:749–750

Xia L, Siemann M, Delius JD (2000) Matching of numerical symbols with number of responses by pigeons. Anim Cogn 3(1):35–43

MIX
Papier aus verantwortungsvollen Quellen
Paper from responsible sources
FSC® C105338

If you have any concerns about our products,
you can contact us on
ProductSafety@springernature.com

In case Publisher is established outside the EU,
the EU authorized representative is:
Springer Nature Customer Service Center GmbH
Europaplatz 3, 69115 Heidelberg, Germany

Printed by Libri Plureos GmbH
in Hamburg, Germany